Developing
Number Concepts
Using Unifix® Cubes

Developing Number Concepts Using Unifix® Cubes

Kathy Richardson

Addison-Wesley Publishing Company

Menlo Park, California • Reading, Massachusetts
London • Amsterdam • Don Mills, Ontario • Sydney

Art Director and Design: Jill Casty and Company
Illustrator: Jeannie Brunnick

ISBN-0-201-06117-1
  JKL-ML-8909

Dedicated to the memory of

Mary Baratta-Lorton

"…they say love is touching souls. Surely you touched
mine cause part of you pours out of me in these lines
from time to time."   (From "A Case of You", Joni Mitchell)

Love and special thanks

…for all your help and support
  to Marilyn Burns, Janann Roodzant, Lynn Pruzan,
  Karen Morelli, Margie Gonzales, Concha Cesena,
  and Cecelia Huerta
…for patient anticipation
  to my husband Tom
  and my children, Susan, Laura, and Todd
…for your belief in me
  to my parents, Jim and Alice Young
  and to Bea Baratta

Introduction

This book is for teachers who want to make mathematics understandable for their students. The concepts dealt with range from beginning counting to the introduction of multiplication and division. The book is based on the premise that children are taught mathematics so it can serve as a useful, problem-solving tool in their lives.

Children's first experiences with numbers influence the way they will deal with mathematics the rest of their lives. Mathematics will be useful for children if we teach it in ways that make sense to them. In order to make sense of number concepts, children need experiences with real things rather than with symbols. There is no meaning inherent in a symbol. When we try to teach mathematics to children primarily through symbols, children learn to treat the symbols as though they were the reality, and true understanding of number concepts is hindered. Understanding is enhanced, however, when children are given opportunities to first learn a concept using real things and then are taught to label that concept with the appropriate symbol(s).

The Unifix Cubes

Unifix cubes were developed in response to children's need to encounter number ideas and relationships in the real world. They are brightly colored plastic cubes that can be snapped together. The Unifix cubes can be used effectively to represent a wide range of number concepts, because they can be counted one by one or joined together to form lengths of any number. The use of various colors can also highlight particular number combinations and relationships. This book was written to help you make effective use of the cubes when teaching number concepts to your students.

How the Book Is Organized

Developing Number Concepts Using Unifix Cubes is organized around the basic concepts usually taught in elementary schools. Each chapter begins with a brief explanation stating what you need to know about the particular concept being presented. A variety of activities designed to give children experiences with the concept are then described. These activities are followed by a group of activities that will help children connect symbols to the concepts already learned.

Children gain understanding of a concept not through one or two well-planned, carefully sequenced lessons but through many and varied experiences with the concept. Two kinds of activities are presented here: teacher-directed activities and independent activities.

TEACHER-DIRECTED ACTIVITIES

The teacher-directed activities are designed for a small group of children who have similar needs. You will lead the children through a variety of experiences, helping them to focus on the concept being presented. It is very important that you provide a variety of activities so that you can help children see and experience the same concept from many perspectives.

INDEPENDENT ACTIVITIES

The independent activities are those that the children can do without interaction with you. These activities are very important, because much learning takes place when a child is confronted with a task to be worked out on his or her own. Carefully chosen independent activities can not only free you to teach those who need you but also can give the children much greater understanding than is possible if you are always there to guide and direct their thinking.

At the end of each chapter are guidelines to help you analyze and assess your children's needs. When making instructional decisions, knowing the level of understanding your children have reached is much more helpful than knowing only what answers they have learned to say or write. This section is intended to help you become more aware of the clues your children will give you so that you can observe the growth of their understanding of numbers.

Using the Book with Your Children

INTRODUCING THE UNIFIX CUBES TO YOUR CHILDREN

Imagine for a moment that someone opened a box and poured out in front of you something you have never seen before—something bright, colorful, and very intriguing. You would naturally be drawn to that material, wanting to touch it, explore it, and see how it works.

That very reaction is the reaction you can expect from your children when they see the Unifix cubes. They will be interested and intrigued—and delighted to find they can snap the cubes together. When their minds are filled with their own ideas and the curiosity of exploring the cubes, it will be hard for them to focus on any specific task that you may set for them. Allow children time to explore the cubes freely before you present them with a directed task. They will need time to satisfy their need to explore the cubes—not just for a few minutes, but for several days. When the intensity of their involvement lessens and they seem to be looking for more ideas for things to do with the cubes, they will be ready for the directed lessons you have in mind for them.

The children's exploration is extremely beneficial to them. They will be naturally estimating, counting, sorting, and comparing lengths in a setting that requires cooperation. You will see children snapping cubes together, eager to see how long they can make their trains. They will be counting and measuring their trains to see if they are longer or shorter than their friends' trains, the rug, or bookshelf. They will join their trains together to see if they can make a train that reaches across the whole room. They will be sorting by color and arranging trains in patterns. They will be inventing all sorts of things—buildings, fences, forts, and various creatures.

Relax and allow the children the time they need to explore the Unifix cubes and the time you need to observe their explorations before you begin using the activities in this book.

After you begin teaching the activities, continue to provide opportunities all year for children to work freely with the Unifix cubes during a free-choice time. They will be more able to focus on the tasks you assign if they know they will have other times to explore their own ideas.

TEACHING THE CONCEPTS TO YOUR CHILDREN

Become familiar with the chapter or chapters that deal with the concepts your students need to learn. Plan a schedule so that you provide children with opportunities for both teacher-directed and independent experiences.

PLANNING THE TEACHER-DIRECTED ACTIVITIES

Group children so that those with similar needs can work together. A twenty-minute instructional period is usually a sufficient length of time to work with one group. Sometimes you will work on one activity during this time, but most of the time you will present three or four short activities. Not all children in your class will need equal time with you. Some groups will need to meet with you briefly every few days and will be able to accomplish much by working independently. Other children will need more teacher guidance to focus on the concepts and will need to meet with you more often.

USING THE INDEPENDENT ACTIVITIES

Children usually will not need to be grouped when working with the independent activities; children of all levels can work side by side. This is possible because many of the independent activities are open-ended and can be experienced at a variety of levels. Other activities can meet individual needs, because the size of the number to be worked with can be varied according to the needs of the children. For example, two children working on addition and subtraction concepts can be working with the same activity at the same time; one child can be learning combinations for three and the other for eight.

Sometimes, it will be appropriate to assign particular activities to particular children, and other times you will want to allow children to choose among many activities dealing with the same concept. When the activities all deal with the same concept, the children will be learning what they need to know no matter which activity they choose.

Be sure you have carefully introduced each activity before you ask children to work with it by themselves. Many activities will need just a brief explanation. Others, however, will need to be modeled several times before the children will know how to play them. Rather than spending a great deal of time teaching the same game over and over to several small groups, you can gather the whole class around you (or half the class if you do not want to work with very large groups) and follow these steps: (1) Play the game in front of the children, using one child as a partner if a partner is needed. Describe each step as you go. (2) Repeat the game, but this time have the children tell you each step. Do not expect any one child to know all the steps yet, but what one child forgets another will remember. (3) Spend a few minutes at the beginning of the math period for the next two or three days having the children tell you what to do at each step. (4) When many of the children understand the game enough to help those who don't, put it out at independent activity time.

The independent activities are important not only because they keep children productively engaged in tasks and free you to teach others, but also because it is vital that children learn to work independently without being guided step by step by a teacher. You will learn much about your children if you arrange your schedule so that you can sometimes observe your children as they work on these independent tasks. On occasion, you will also find it appropriate to pose challenges and give help if needed.

Using the Textbook

When you are making decisions about your math instruction, keep in mind that a textbook, by its nature, can present only symbolic representations of concepts—not the concepts themselves. Children learn concepts from experiences with real objects. They cannot learn concepts from symbols in a textbook. They must already know a concept and how it is symbolized *before* they can work in a textbook with real understanding.

It is possible for children to learn to fill in answers in a textbook without understanding what they are doing. They memorize the rules for doing the page and get answers by manipulating symbols. Memorizing rules hinders rather than enhances concept development. It stops children's search for the underlying sense of things and teaches them that math is something it is not—black marks on paper.

Examine your textbook to determine the concepts you are expected to teach. Teach those concepts and the related symbols before asking children to work with those concepts in the textbook. View the textbook as material that provides symbolic reinforcement for already understood concepts.

Acknowledgements

Developing Number Concepts Using Unifix Cubes is an outgrowth of the author's extensive personal interaction and involvement with Mary Baratta-Lorton and the Center for Innovation in Education, Inc., co-founded by Mary and Bob Baratta-Lorton, which has generated the following Addison-Wesley publications: *Workjobs, Mathematics* THEIR *Way, Math…A Way of Thinking*, and *Workjobs II*. The author has worked with teachers and children for many years using the philosophy and approach of these publications. Elaborations and adaptations based on her personal experiences with children and their developing mathematical ideas resulted in the writing of *Developing Number Concepts Using Unifix Cubes*. The theme and thread especially of *Mathematics* THEIR *Way* and *Workjobs II* are present throughout this book, so it is difficult to be specific in giving credit and acknowledgement. We have indicated by footnote those places where such credit can be precisely shown.

Use of the counting boards throughout the book is an adaptation of *Workjobs II* format and sequence of skills.

Permission has been granted for the use of these ideas and activities by Addison-Wesley and the Center for Innovation in Education, Inc.

Grade Level Helps

Even within the same grade level, the needs of children vary greatly from one class to another. View the following outline as a guide for a place to start with the grade you work with. After you have dealt with the ideas in this book for a while, you will be able to look to your children, not to this outline, for the final word on what is appropriate for them.

KINDERGARTEN

Chapter One: Beginning Number Concepts

Chapter Two: Pattern
Emphasis on Section I: Developing the Concept of Pattern

Chapter Three: The Concepts of More and Less
Emphasis on Section I: Developing the Concepts of More and Less

Chapter Four: Beginning Addition and Subtraction
Emphasis on exploring number, as in Counting Boards—Making
Up Stories, p. 87
Number Trains, p. 95
Number Arrangements, P. 97
Number Shapes, P. 99

FIRST GRADE

Focus on

Chapter Four: Beginning Addition and Subtraction

Chapter Two: Pattern
Emphasis on Section I: Developing the Concept of Pattern

Chapter Three: The Concepts of More and Less

Chapter Five: Place Value
Emphasis on Section I: Developing the Concept of Regrouping
Section II: Developing a Sense of Quantities above
Ten

For children who need extra help:

Chapter One: Beginning Number Concepts

For children ready for more:

Chapter Five: Place Value
Section III: Addition and Subtraction

SECOND GRADE

Chapter Five: Place Value

Chapter Two: Pattern

For children still insecure with number relationships, basic facts, and word problems:

Chapter Three: The Concepts of More and Less
Emphasis on "How many more or less?"

Chapter Four: Beginning Addition and Subtraction

For children ready for more:

Chapter Six: Beginning Multiplication

Chapter Seven: Beginning Division

THIRD GRADE AND ABOVE

Focus on

Chapter Five: Place Value

Chapter Two: Pattern
Emphasis on number patterns

Chapter Six: Beginning Multiplication

Chapter Seven: Beginning Division

For children insecure with number relationships, basic facts, and word problems, see the following chapters for ideas to help you:

Chapter Three: The Concepts of More and Less

Chapter Four: Beginning Addition and Subtraction

Contents

CHAPTER ONE

If your textbook or workbook objectives
are:
- Matching sets
- Counting
- Matching sets and numerals

Then you are dealing with:

Beginning Number Concepts

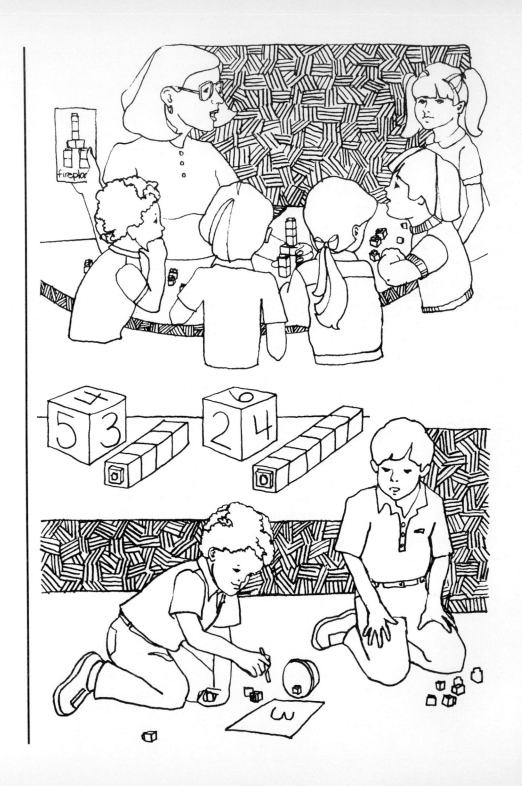

WHAT YOU NEED TO KNOW ABOUT BEGINNING NUMBER CONCEPTS

When a young child says, "One, two, three, four, five . . ." and responds correctly when asked to name the symbols for those numbers, parents and teachers are pleased. Many feel the child is well on the way to "knowing her or his numbers." However, developing the concept of number is much more than knowing how to say number names in correct order (counting by rote). It is much more than saying the appropriate words when shown squiggles in the shapes of numerals (recognizing numerals). Developing the concept of number requires not just memorizing words and symbols but also thinking about things in a special way.

Much of young children's time and energy are spent learning the names of the things in their world. They learn ball and water and apple. They are encouraged to notice how things look and they learn round and wet and red. But when dealing with the concept of number, children must ignore those physical properties they have spent so much time paying attention to, for number can't be seen. Number is an idea.

When we analyze the concepts that young children need to understand in order to work successfully with number ideas, we find they are quite complex. Three of the key concepts children need to develop are inclusion, one-to-one correspondence, and conservation of number. We can work with children with much more understanding if we are aware of these basic notions.

Inclusion. A young child who has not developed the idea of inclusion when counting oranges may respond to the command "Bring me three" by picking up the orange the speaker was pointing to when he or she said "three." The child is naming the "three" orange in the same way she labeled "book" or "chair." What she must learn is that "three" is not the name of a particular orange. In fact, the label "three" *includes* the oranges labeled "one" and "two" as well. She must also become comfortable with the fact that in recounting those same oranges, the orange she labeled "three" could just as well have been labeled "one" or "two." She must realize that although it does matter in which order you say the words (saying "one, two, three" is all right but saying, "two, three, one" is not), it doesn't matter in which order you point to the oranges as you count them.

One-to-One Correspondence. Young children who are just discovering the idea of counting notice that people point to objects while they say number words. Children who are still not secure with the idea will say

words faster or slower than they point, having noticed only that they must stop pointing and stop saying words at the same time. In this process they may skip or may recount any object. It takes a while for children to recognize that you must say (or think) one word for each object. This need to match one word with one object is referred to as *one-to-one correspondence*. Children develop this skill gradually. A child who has developed one-to-one correspondence to eight or ten may lose track of the notion when working with larger numbers.

Conservation of Number. Children's perception of the world (i.e., what they *see*) plays an important part in their understanding of the world. What *appears* to be true is easier to believe than what would seem logical from an adult point of view. At a certain stage a child believes that if something looks different, it *is* different. It is natural, then, for a child who sees several objects lined up to think there are more when the same objects are spread out. It is natural to think there is more cracker when one is broken into several pieces than when it is whole. It is natural to think that five elephants are more than five peanuts. It is often difficult for adults to understand how children think when they see a child count out six objects and then say that there are more than six when the objects are spread out. It may be easier for us to understand if we think about those situations where we are also fooled by our perceptions. Thirty adults in a room seems like more than thirty children in that same room. If you don't actually count, your estimation might even reflect that impression. Our experiences over long periods of time have taught us to check our perceptions and trust our logic when perceptions and logic are contradictory. Children are still tied strongly to their perceptions. Knowing that the number of objects does not change when the objects are moved, rearranged, or hidden is called *conservation of number*.

Rote counting and numeral recognition are useful skills for young children to develop. However, practicing saying a series of number words in correct order and learning to name the numerals do little toward helping children develop meaningful number concepts. The activities in this chapter provide children many opportunities to count real objects in a variety of ways. It is through this kind of experience that children develop not only counting skills but also a strong sense of number. Symbols come to have meaning for the children because they are introduced not as ends in themselves but as labels for groups of quantities.

Section I: Developing Beginning Number Concepts

The activities in this section focus on the skill of counting, with special emphasis on developing the notion of one-to-one correspondence. Although the activities allow children opportunities to practice counting, they also allow children to match and compare quantities. Counting skills are developed along with ideas such as six is one more than five, eight is less than nine, and seven remains seven no matter how the cubes are arranged. The emphasis in this section is on developing meaningful number concepts, and no written symbols are used at this stage.

Teacher-Directed Activities

The activities are designed to be done with a small group of six to eight children with similar needs. (See p. 23 for help in determining which children belong in a particular group.) Three to five of the activities can be done one after the other during a fifteen- to twenty-minute instructional period. Each day play one or two familiar games with the children and introduce one or two new ones. When all the activities are familar to the children, choose from any of them. There is no particular sequence that should be followed.

Each activity should be repeated many, many times. Only after children know how to do a particular activity will they be able to focus on the concept the activity is designed to teach. Most of the activities can be played at a variety of levels simply by changing the size of the number(s) being worked with. Even with much repetition, children's interest will be maintained if you provide a variety of activities within an instructional period, if you choose freely from the activities in the chapter, and if the size of the number being worked with is appropriate.

SLIDE AND CHECK*

Materials: Unifix cubes • Xylophone (optional)

The children slide cubes one at a time toward themselves as they practice counting with you over and over to a designated number. One group of children may need practice counting to five; another group

of children may need to count to nine. Play a note on the xylophone on each count.

For example: We are going to practice counting to four.

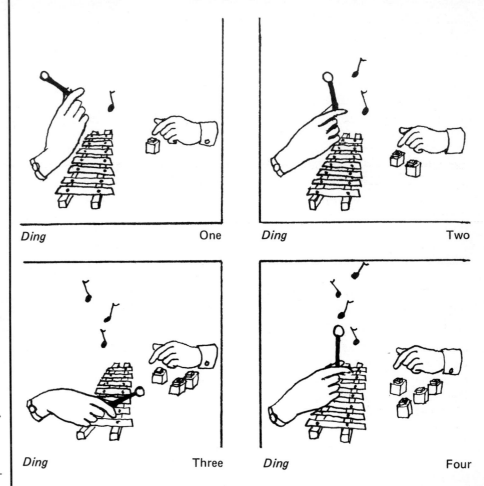

Ding   One   Ding   Two

Ding   Three   Ding   Four

After the number has been counted, say "Check," and have the children recount the cubes.

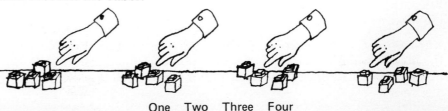

One   Two   Three   Four

Run the stick across the xylophone as a signal to push the cubes back into the pile.

Repeat several times.

Variation: If no xylophone is available, just count with the children as they slide cubes and say "Push them back" when they are to return the cubes to the pile.

COUNT AND DUMP

Materials: Unifix cubes • 1 margarine tub per child

The children each drop Unifix cubes into their margarine tubs as they practice counting to a designated number. You or a student say "Dump," and the cubes are dumped and recounted.

For example:

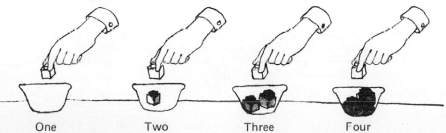

One   Two   Three   Four

Dump

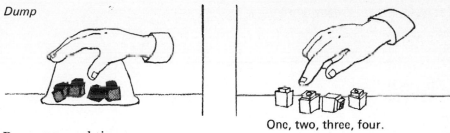

One, two, three, four.

Repeat several times.

THE ONE MORE COUNTING GAME

Materials: Unifix cubes

Play the following game with the children, having them count as high as they need to practice.

Get one cube
How many do we have?

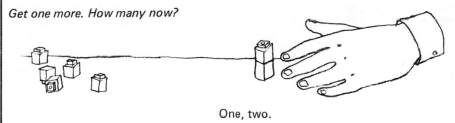

One.

Get one more. How many now?

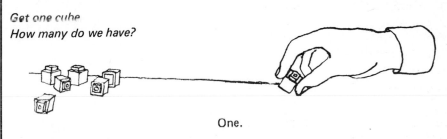

One, two.

Get one more. How many now?

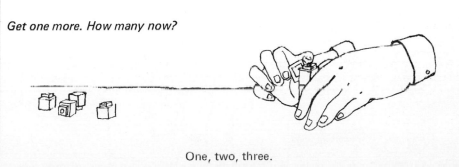

One, two, three.

Get one more. How many now?

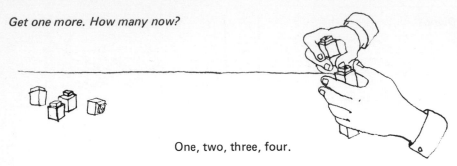

One, two, three, four.

When the children have counted as high as you wanted them to, have them build another stack. Build several stacks with them so they can get a lot of practice with the sequence they need to work on.

CREATIONS

Materials: Unifix cubes • Creation cards—see p. 213 for directions for making

Give the children practice with one-to-one correspondence and matching by having them build Unifix creations to match your actual model or the picture on the creation cards. (When introducing the cards, show the children how to build the creation standing up rather than lying flat.) Require the children to make it exactly like the model or card, using the correct number of cubes for each part of the creation.

PEEK AND COUNT

Materials: Unifix cubes • 5–6 margarine tubs

Have the children close their eyes while you hide sets of cubes under the margarine tubs. When you say "Peek," a child lifts a tub and points to the cubes while the other children count together to see how many were hiding. After all the sets have been counted, the tubs are rearranged (reminiscent of the old shell game), and the game is repeated.

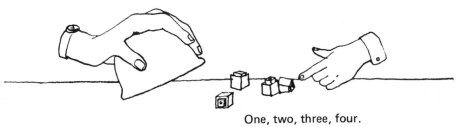

One, two, three, four.

HUNT FOR IT

Materials: Unifix cubes • 5–6 margarine tubs

Hide various sets of cubes under five or six margarine tubs. The children take turns lifting the tubs so the other children can peek and look for a specific number. Each time a tub is lifted, they tell if they have found the number they were looking for. If not, they tell if the set of cubes is too much (more) or not enough (less). Have the group respond rather than an individual child. This should not be a testing situation but simply a motivator to count and compare.

For example:

We are looking for four.

Nope, that's five. That's too much.

HIDE IT

Materials: Unifix cubes • 1 margarine tub

Use just one margarine tub. Let each child take a turn hiding cubes under the tub while the other children close their eyes. To limit the number of cubes the child can hide, make available the maximum number of cubes you want the children to count. For example, if you do not want your group to count past six, give each child only six cubes to use. The child takes as many of those cubes as she or he wishes to hide and then says "Peek," lifting the tub so the other children can count the cubes.

FIND A MATCH*

Materials: Unifix cubes • 6–10 margarine tubs

This game is a form of "Concentration," in which children look for matching sets of cubes. Place Unifix cubes under margarine tubs so that there are two of each number you want the children to work with. Arrange the tubs in rows and columns. Have the children take turns lifting one tub and then another to see if the sets of cubes match. If the cubes match (i.e., if they are the same number), the child removes the cubes and the tubs. If not, he or she replaces the cubes under the tubs and the next child takes a turn trying to find pairs of numbers that match.

Variations: Match loose cubes with sets of joined cubes.

Or, put cubes under half the tubs and matching cards with dots under the other half. (The dots should be drawn in dice or domino patterns, so the children can begin recognizing them.)

Or, allow the children to keep looking until they find the matching set instead of lifting only two tubs at each turn.

COUNTING STORIES

Materials: Unifix cubes separated into the various colors • Working space papers —see p. 210 for directions for making or • Counting boards representing specific settings such as corrals, fields, trees, oceans, etc.—see p. 210 for directions for making

Tell stories and have the children act them out, pretending the Unifix cubes are the objects in the stories. They put the Unifix cubes on the counting boards or working space papers, which represent the setting for the story.

For example, using working space papers:

Three children are playing on the slide. Two children are playing in the sandbox. Count the children.

*Based on MATHEMATICS *THEIR* WAY, p. 191.

A mama bear and a daddy bear are walking in the woods with their two babies. Count the bears.

For example, using counting boards:

Four trucks are driving on the freeway. Two cars pass them.

Three blue motorcycles were on the road. Two red motorcycles were on the road, too.

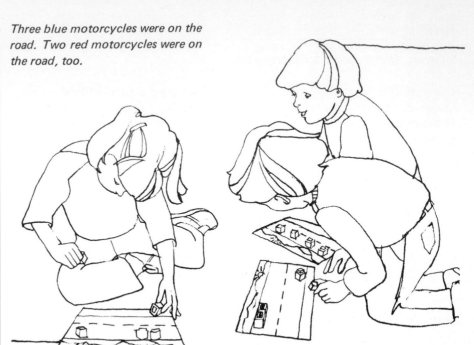

This activity gives children a chance to develop language as well as number concepts. Encourage lots of discussion and use of imagination. Allow the children also to make up stories. Most children at this level will not be concerned with "How many altogether?" or "How many left?"–just counting out the appropriate groups is enough.

Note: Using cubes to represent objects presents an opportunity to discuss what color cubes would be appropriate. Could there be an orange truck? An orange bear? What color could the bear be? It is helpful to have the cubes sorted into the various colors. After a discussion of possible colors, the appropriate color(s) can be made available to the children. If you do not want to deal with the subject of color, make up stories in which the color of objects doesn't matter, or simply accept any cube to represent any object, whether or not it is realistic.

FINGER COUNTING

Materials: Unifix cubes

Say a number. Have the children put the appropriate number of cubes on their fingers using one or both hands. Repeat the same number three or four times, asking them to find different ways to show that particular amount.

Four.

Show me a different way.

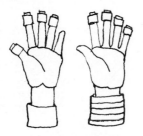 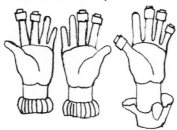

BREAK IT UP

Materials: Unifix cubes • 1 margarine tub per child

Say a number and have the children build the appropriate Unifix train. When you say "Break it up," the children break their trains into their tubs. They guess how many they have in their tubs, then count and check. This may seem too obvious to you, but you will learn that, for many children, it is not at all obvious that the same number of cubes will remain after breaking them apart. Repeat several times using various numbers.

GIVE AND TAKE

Materials: Unifix cubes • 1 margarine tub per child

Start with a set of loose cubes, which the group will count together as the teacher or student points.

One, two, three, four.

Cover the cubes with a margarine tub.

Without uncovering the cubes, lift the tub slightly and put another cube under the tub. Have the children guess how many are under the tub.

How many?

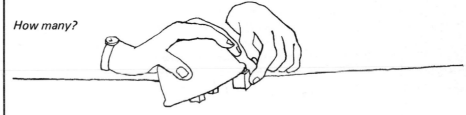

Five, seven.

Lift the tub to check.

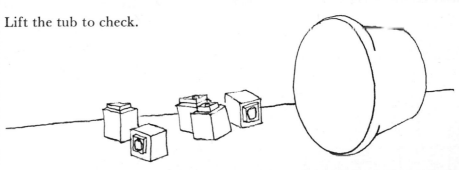

One, two, three, four, five.

Continue the game by adding cubes under the tub or taking cubes away one at a time. Each time a cube is added or taken away, have the children guess the number and then check.

Extension: Put two cubes under the tub or take two cubes away each time.

Note: Some children will have no idea each time you add or remove a cube how many will be hiding, and their guesses will reflect that. At its simplest level, then, this game is a fun way to practice counting. It also provides the opportunity for children to develop visual images of numbers, and it encourages using the skills of counting on and counting back. Observe your children as they guess, and accept whatever level they are able to operate on.

TELL ME FAST

Materials: Dot pattern cards with 2–6 dots (teacher-made; see examples below) • Unifix cubes

Show the children some cards with patterns of dots arranged on them. Have them take the number of cubes they think are required to cover each dot on a card. Then have one child put cubes on the dots while the group counts and checks. After the children have become familiar with the cards, hold a card up briefly and put it down quickly to encourage the children to try recognizing the number of dots without counting one by one.

There are five dots.

GRAB BAG COUNTING

Materials: Unifix cubes (1 color) • Paper bag

The children take turns grabbing a handful of cubes. Each time a handful is taken, the child places the cubes in front of the group so they can count together to determine how many cubes were grabbed. (To get larger numbers than some small hands can grab, use a margarine tub as a scoop, or have the children grab two handfuls.)

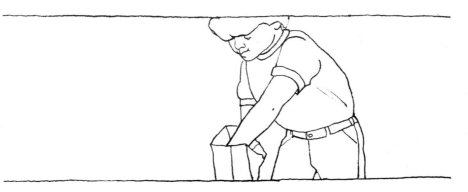

GROW AND SHRINK*

Materials: Unifix cubes • Working space paper—see p. 210 for directions for making

This activity is a simple counting game, but one that allows children to begin to see relationships between numbers.

    Say a number.   *Four.*

The children place that many cubes on their working space paper (one cube on each dot).

*Based on MATHEMATICS *THEIR* WAY, p. 190.

Say another number and have children build that number.

Seven

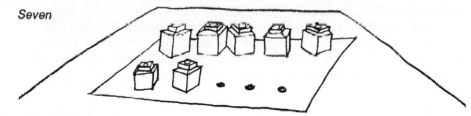

Continue saying numbers.

Three.

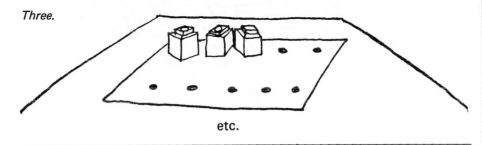

etc.

Note: You will notice that children approach this task in different ways. Some children remove all the cubes each time and then build the assigned number. Other children add cubes or take cubes off as needed to make each new number. (That is, if they have four already on their paper, they will get two more to make six rather than starting over each time. If six cubes are already on the paper and they are to make four, they will take two off.) Do not attempt to teach the children to do it the better or easier way. Simply observe them and notice the level of thinking at which they are presently operating.

Variation: Let the children take turns rolling a die to indicate which number to build. A large dotted die can be made from a milk carton—see p. 215 for directions for making.

TALL AND SHORT

Materials: Unifix cubes • Large milk-carton dotted die—see p. 215 for directions for making

Have the children take turns rolling the large die and then have them build a tower as indicated by the die. Continue to roll the die and add

or take cubes off the Unifix tower. Notice which children are able to add cubes or take them off without counting each time and which children need to count.

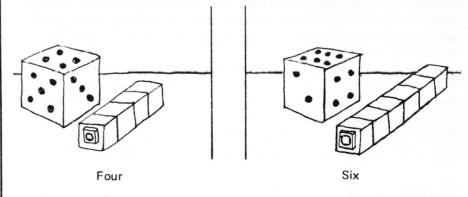

Four   Six

Note: This is very similar to "Grow and Shrink." Varying an activity even slightly provides an additional opportunity to experience the concept. As in "Grow and Shrink," do not try to teach the children how to play the game efficiently. It is enough to provide the opportunity for the children to see number relationships when they are ready. For the children, learning to do what the teacher wants before they are ready does not really lead to deeper understanding.

Independent Activities

The following activities are designed to be done by children working independently while you are teaching another group. This assumes the activities have been carefully introduced during a directed instructional time.

The children do not need to be grouped by ability in order to perform the independent activities. Individual needs can be met with children of many levels working side by side. Some activities require that children know what specific number they are to work with. For example, one child may need to be counting out sets of three onto counting boards, while another child may need to be counting out

Variation: Start with a train of a particular length, and break cubes off according to the dice.

BUILD A CITY*

Materials: Unifix cubes • "Build a City" game board (run off on construction paper using black-line master 33) • Dotted dice (a variety of dice can be used, depending on the numbers you want the children to build—see p. 215) • More/less spinner—see directions for making, p. 217

Children take turns rolling the dice and placing Unifix buildings on their side of the game board. When all the spaces for buildings have been filled, the children snap their cubes together and compare them. They then turn the spinner to see if the person with more or less cubes is the winner. (This game is especially effective for providing children the opportunity to deal with the concept of zero. If the child rolls zero dots, she or he leaves a space empty.)

CREATIONS

Materials: Unifix cubes • Creation cards—see p. 213

The children copy the creation cards using Unifix cubes, as shown on page 4. They must use the exact number of cubes shown on the cards.

Section II: Connecting Symbols to the Concept

This section deals with numeral recognition and the association of numerals with the appropriate quantities. A most basic idea dealt with is the idea of symbolization itself. All activities presented here encourage the idea that the symbols do not exist in and of themselves but that they only *represent* real things. It is not unusual for a young child to have learned the names of the numerals before she or he has any idea of what those squiggles are about, just as an adult can learn to say and write the symbol πr^2 without any real understanding of the concept represented by that symbol. Knowing a symbol without knowing the concept represented by that symbol is virtually useless; therefore, the numerals presented in this section are never dealt with in isolation but always in association with the quantity represented. The goal is for the symbols to trigger in the child's mind visual images of what they stand for.

When you introduce your children to numerals, you will continue to play the familiar games, simply adding symbols to them. In this context children continue to develop and strengthen their number concepts.

Teacher-Directed Activities

When the children have a strong base of counting skills (i.e., rote sequence and one-to-one correspondence to ten), they are ready to begin associating the numerals 0 to 9 with appropriate quantities. When you feel you have a group of children who are ready, introduce them to the symbols by modeling the numerals for them. Whenever you say a number, write it as well. Replace dotted dice with numbered dice. In general, give the children opportunities to see and hear the names of numerals in association with the appropriate sets. In the beginning, read the numeral cards, the numbered dice, or the numerals

*Based on MATHEMATICS *THEIR* WAY, p. 320.

you have written. Soon the children will be reading the numerals along with you. Do not overwhelm them with all ten numerals at once. Let children become confident and at ease with a few symbols at a time. Of course, some groups will be able to deal with many symbols quickly, whereas others will need lots of practice with a few symbols for some time.

All the following activities allow children the opportunity to learn to associate numerals and quantities without the extra burden of writing the numerals. Either the teacher does the writing or the children use dice or numeral cards.

COUNTING STORIES (Introduced on p. 5)

Materials: Unifix cubes • Counting boards or working space papers

As you tell children stories to be acted out using the counting boards or working space papers, write the numerals as you say them.

For example: There were three (write 3 on the chalkboard as you are saying it) cars in the parking lot at the (local) shopping center. Two more (write 2) cars came to the shopping center. Four (write 4) cars drove away.

When the children become familiar with the symbols, write the numeral without saying it first to allow the children the chance to see if they can remember it.

For example: This many (write a numeral) children were playing on the playground. How many children did I write? (Some children will be able to respond correctly. Do not call on individual children at this point. This is to be a teaching situation—not a testing situation. Those who are not sure of the numerals yet will need more time hearing you and the other children say the names of the numerals.)

GROW AND SHRINK (Introduced on p. 8)

Materials: Unifix cubes • Working space papers • Large numbered die—see p. 215 for directions for making

Have the children take turns rolling a large, numbered die. (See directions for making on p. 215.) Each child will then build the number rolled on their working space paper. For some groups of children you may use a die that has only the numerals 0 to 4; for another group,

you may use a die with numerals 4 to 9. (Each time a number is rolled, you and the children who are able call out the number rolled. The more a child who doesn't know the numerals sees and hears the names of the numerals, the sooner he or she will learn them.)

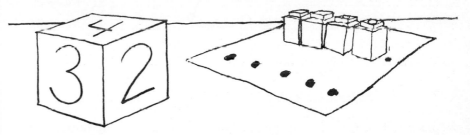

HUNT FOR IT (Introduced on p. 4)

Materials: Unifix cubes • 5-6 margarine tubs • Large numbered die

Hide sets of Unifix cubes under the margarine tubs to match the numbered die you are using. Roll the die to determine what number the children are to look for. The children take turns lifting the tubs so the group can count the cubes to see if they have found the number rolled.

FIND A MATCH (Introduced on p. 5)

Materials: Unifix cubes • 6-10 margarine tubs • Numeral cards—see page 214 for directions for making

Instead of hiding two sets of cubes, hide numeral cards and matching sets of cubes under the margarine tubs. The children take turns looking for the matching sets and numerals.

TALL AND SHORT (Introduced on p. 9)

Materials: Unifix cubes • Large numbered dice

Replace dotted dice with numbered dice to determine the numbers to be built.

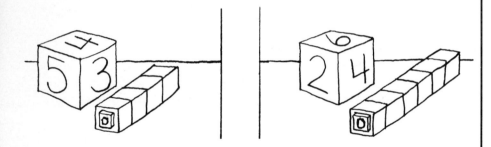

Independent Activities

When children are learning numerals, allow a proportionally larger percentage of time for independent activities as compared with teacher-directed activities. Children working independently with the symbols are forced to notice how they look and to distinguish between them. This is an effective way to learn, as long as children are provided with the means for finding out what the numerals are if they forget.

Individual differences can be taken care of and children of many levels can work side by side if you provide the following kinds of help for children who need it. Put up number cards in the classroom that can be seen easily by all the children. Each numeral should be large and have next to it the number of dots (arranged in dice or domino patterns) that indicate what the numeral is.

Small number lines (from 1–10) that the children can get when they need them should also be available. The numeral cards used for some of the activities have the appropriate number of dots on them so the children can figure out the numeral when they forget. Numerals only are on the reverse so children can work without the help of the dots when they are ready.

Have a variety of dice available also. You can color code the dice so that you can distinguish between the dice with smaller numbers and the ones with larger numbers. (See notes on making and using dice on p. 215.)

It is likely that you will continue to have some children who are not ready for symbols when others are. You can meet these individual needs by including in an activity both dotted and numbered dice and by allowing some children to count out cubes onto counting boards according to their dot cards while other children use numeral cards to indicate the appropriate number of cubes. There is no need for children who are still working with dotted dice or dot cards to feel inadequate. All tasks can be valued and accepted as worthy endeavors as long as the appropriate effort is being made. Playing the same games at different levels does much to promote acceptance of a variety of tasks by children working together.

COUNTING BOARDS (Introduced on p. 5)

Materials: Unifix cubes • Numeral cards—see p. 214 for directions for making • Counting boards • Containers (margarine tubs) for numeral cards or • Envelopes for each child

The counting boards give children an excellent opportunity to practice reading numerals and counting out sets to match. They will spread out the set of counting boards they have chosen, place a numeral on each board, and count out the appropriate cubes. You can tell the children to take numerals they will be using from a container that holds numerals from 0 to 5 or, if appropriate, from another container that holds numerals from 0–9.

If you wish to be more specific in the assignment of numerals, provide children with envelopes containing the numeral cards you have determined each needs to work with. Some children will have an envelope of 1's, 2's, and 3's. Other children may have numerals to six. If you want children to learn the numeral 5, have their envelopes contain several 5's. Be sure to include other numerals that the children already know so they will be forced to look carefully at each one to determine what it is. When you decide some children need additional numerals to work with, add them to their envelopes.

UNIFIX PUZZLES

Materials: Unifix puzzle cards—see p. 211 for directions for making • Unifix cubes • Numeral cards—see p. 214 for directions for making

The Unifix puzzles are made of various paper shapes. The children fill the Unifix puzzle cards with cubes and determine the number of cubes that fit in each puzzle. They label the puzzles with the tiny numeral cards.

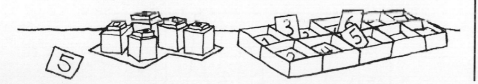

Extension: Unifix Puzzle Graph

Materials: Unifix puzzle cards • Group-size graphing sheet (36" x 48" sheet of paper or plastic marked off in rows or columns) • Unifix cubes

This activity can be done in a small-group, teacher-directed lesson as a follow-up to work with the Unifix puzzles.

The children fill Unifix puzzle cards with Unifix cubes. When they have counted to see how many cubes fit in the puzzle, they place it in the correct column in the graph. They can then notice all the variations for each number.

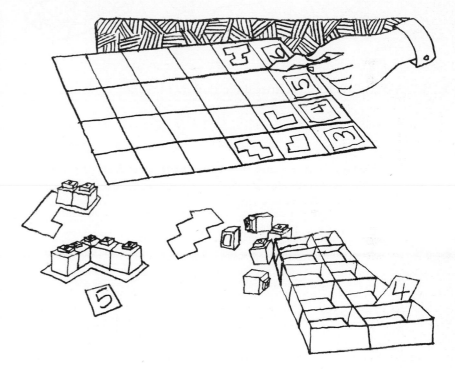

LINE PUZZLES

Materials: Line puzzles—see p. 212 for directions for making Unifix cubes • Numeral cards—see p. 214

The children line up cubes along the line puzzles and label them with the numeral that tells the number of cubes that fit along the lines.

ROLL-A-TOWER RACE (Introduced on p. 11)

Materials: "Roll-a-Tower" game board • Dice that match the numerals on the board

...use the Roll-a-Tower game board has both numerals and dots, it ...be used by children who are just learning the numerals as well ...y children who already are quite confident in recognizing numerals.

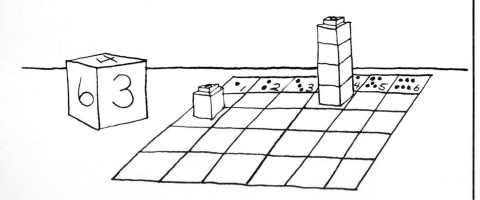

BUILD A CITY (Introduced on p. 12)

Materials: "Build a City" game board • Numbered dice • More/less spinner • Unifix cubes

Replace dotted dice with numbered dice to indicate the height of the buildings.

GRAB BAG COUNTING (Introduced on p. 8)

Materials: Unifix cubes • Paper bag for each child • Numeral cards—see p. 214 or • Unifix number indicator—see p. 217

The children take handfuls of cubes out of a grab bag and label their handfuls with the tiny numeral cards or Unifix Number Indicators or by coloring in squares on black-line master 1.

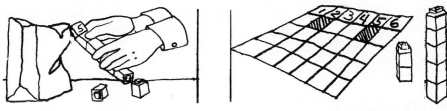

PICK A NUMBER

Materials: Unifix cubes • Paper bag for each child • Numeral cards—see p. 214 or • Unifix number indicator—see p. 217

The children grab for numerals from a grab bag instead of cubes. Have the children build sets to match the numerals they grabbed. (Use numeral cards (see p. 214) or Unifix Number Indicators (see p. 217.))

UNIFIX TRAIN RACE (Introduced on p. 11)

Materials: Unifix cubes · Numbered dice

Use numbered dice instead of dotted dice to indicate the number of cubes to be added to or broken off each partner's train.

BUILD A STAIRCASE (Introduced on p. 10)

Materials: Unifix cubes · Numbered dice

Replace the dotted dice with dice numbered from one to six.

I need to put the three between the one and the four.

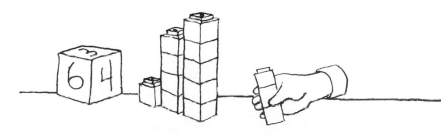

Extension: Provide a commercially available Unifix Staircase or Unifix Value Boats (see p. 217). Use a spinner numbered from one to ten to indicate the steps to build (see p. 216 for directions for making).

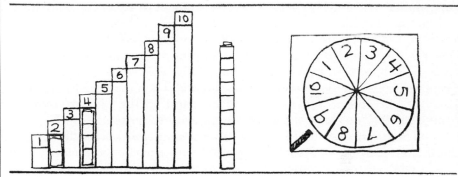

Section III: Writing Symbols to Label the Concept

Asking children to recognize numerals when they see them and asking them to write the numerals involve two different skills. The differences are similar to the differences between reading a word and spelling it. Children who can recognize the word *dog* when they see it have to remember exactly how the word looks in order to write it, and have the physical dexterity to reproduce what they remember. So, a child may be able to read the word *dog* but not be able to write it. Similarly, children may be able to recognize a numeral when they see it, but they may not be able to remember it exactly and, in the case of young children, may not have the physical dexterity to write it.

The physical skill of learning to write numerals is a totally separate task from learning to associate a symbol with a particular quantity. The child who writes the numeral 3 fifteen times will probably become better able to write the numeral 3 but will not know anything more about the concept of "threeness."

We must be very careful not to assume that children are learning anything about quantity and number when they learn the numerals. This is not to say that children do not need practice writing numerals; just be sure you recognize the practice as a handwriting skill rather than a math skill.

Teacher-Directed Activities

It is assumed here that you will provide opportunities for your children to develop the physical skill of writing numerals. Once they have the physical dexterity, they need to *use* this skill to represent quantities so the connection between symbols and the quantities will be made. Play the familiar games, and have the children write the numerals to label the numbers with which they are working.

PEEK AND COUNT

Materials: Unifix cubes • 5–6 margarine tubs • Chalkboard, chalk, eraser for each child

Hide various quantities of cubes under margarine tubs. Lift the tubs one at a time, and have the children write the number of cubes under each tub. Rearrange the tubs (similar to the old shell game) and repeat.

Variations:

One More

Materials: Unifix cubes • 5-6 margarine tubs • Chalkboard, chalk, eraser for each child

Hide various quantities of cubes under the tubs.
Lift the tubs one at a time and have the children write the number of cubes under each tub on individual chalkboards.

Then life the tub slightly, and place one more cube underneath. Have the children predict the number they think will be under the tub now and write that number on their chalkboards.

Lift the tub and have the children count to see how many. If they made a mistake, have them erase and write the correct number.

1, 2, 5, 4, 5, 6
That's what I wrote.

One Less

Continue for all the other tubs.

Materials: Unifix cubes • 5-6 margarine tubs • Chalkboard, chalk, eraser for each child

Play the game as described above in "One More," but in this game remove one cube each time.

GIVE AND TAKE (Introduced on p. 7)

Materials: Unifix cubes • One margarine tub • Chalkboard, chalk, eraser for each child

Have the children write the number of cubes they think are under the margarine tubs as you add cubes and take them away.

COUNTING STORIES (Introduced on p. 5)

Materials: Unifix cubes • Chalkboard, chalk, eraser for each child—see p. 215 for directions for making

As you or the other children tell stories, have the children act them out with the cubes then write the numerals that tell how many. If they have small chalkboards, they can place the cubes right on the chalkboard and then write the numerals underneath the sets of cubes.

For example:

There were three cars in the parking lot.

There were five children on the swings.

Independent Activities

The children should be reasonably comfortable writing numerals before doing the activities in this section. The writing of numerals should not be overly frustrating to the point that it interferes with focusing attention on number concepts. The work should be challenging and satisfying. Many kindergarten children who need additional challenges would be better served by working with activities from the next chapter(s) on pattern, comparing, and beginning addition and subtraction rather than by having the task of numeral writing imposed on them.

COUNTING BOARDS (Introduced on p. 10)

Materials: 2 x 3 pieces of paper • Counting boards (1 set per child) • Unifix cubes

Have children count out quantities onto the counting boards. (If you want them to write numbers no higher than nine or ten tell them they can choose the number of cubes they want to work with, but limit the maximum number of cubes they can use.) The children then write the number of cubes there are on each board onto 2″ × 3″ pieces of paper. When finished, they will staple the papers together to make tiny number books.

Variation: Some children would benefit from a transitional stage before they decide for themselves what quantities to put on their boards. Give them some numeral cards (see p. 214), and have them put out the cubes as indicated by the cards. To make tiny number books, they can copy the numerals onto the little papers. This assures the numerals will be made correctly.

HIDE IT—A Game for Partners

Materials: Unifix cubes • Margarine tub • Chalkboard, chalk, eraser

Partner A hides cubes under a margarine tub. Partner B lifts and counts the cubes and writes the number on a small chalkboard that tells how many were hiding. The partners then switch roles.

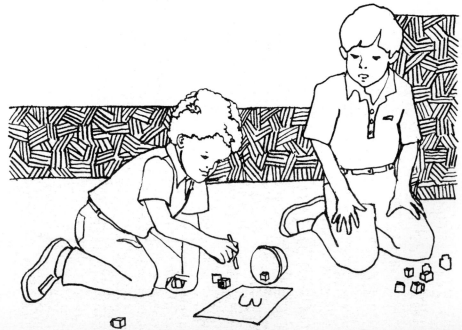

GRAB BAG COUNTING

Materials: Unifix cubes • Paper bag • Worksheets (see black-line master 1 and/or 3)

Each child grabs a handful of cubes from a bag and records the number of cubes he or she got each time. This could be recorded on either of the sheets pictured here.

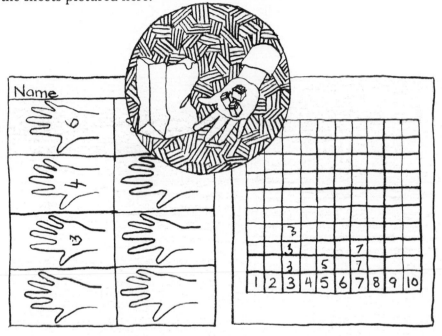

GIVE AND TAKE—A Game for Partners

Materials: Unifix cubes • One margarine tub for each set of partners • Chalkboard, chalk, eraser for each set of partners—see p. 215

Partner A counts out a set of cubes (any number to ten) so that Partner B knows how many cubes there are. Partner A hides the set of cubes under a margarine tub, then reaches under the tub to take one cube away or add an additional cube. Partner B writes on a small chalkboard the numeral that he or she thinks is under the tub. Partner A lifts the tub, and they count and check together. Partner A continues to add or take cubes away and Partner B continues to write for several turns before they switch roles.

UNIFIX PUZZLES

Materials: Unifix puzzles —see p. 211 • Unifix cubes • 2″ × 3″ pieces of paper

Children fill Unifix puzzles with cubes and write that number of cubes on tiny pieces of paper. After doing several puzzles, they staple the papers into tiny number books.

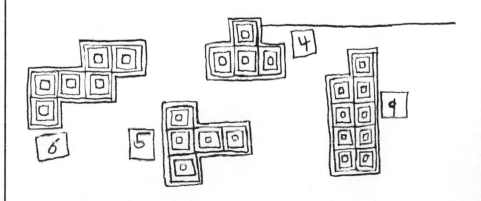

LINE PUZZLES

Materials: Line puzzles—see p. 212 • Unifix cubes • 2″ × 3″ pieces of paper

The children lay Unifix cubes along the line puzzles and then count and write the number of cubes each puzzle took. After doing several puzzles, they staple the papers into tiny number books.

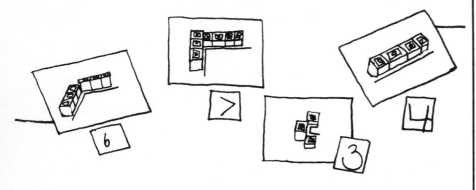

HOW LONG IS IT?

Materials: Unifix cubes • Worksheet (see blackline master 4) • Heavy yarn or ribbon. Be sure the yarn or ribbon is no longer than 10 cubes. Staple a folded piece of tagboard onto the end of each piece of yarn or ribbon. Draw a symbol (colored dots or letters on each piece of tagboard)

Have children measure to see how many Unifix cubes can be snapped together and lined up along pieces of heavy yarn or ribbon, and have them record on the work sheet.

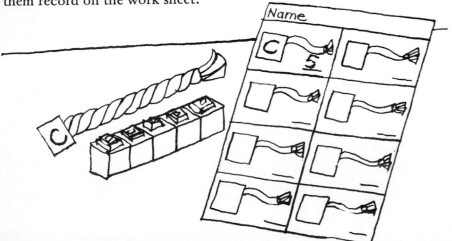

HOW MANY DOES IT HOLD?

Materials: Jars, scoops, margarine tubs, or other small containers • Unifix cubes • Worksheets (see black-line master 2)

The children count the cubes in a variety of small containers and record the numbers. The containers can be labeled with colored dots, simple shapes, or letters, and the following recording sheet can be used.

The child writes the symbol of the container and the number of cubes it contained.

SORT AND COUNT

Materials: Margarine tubs (1 per child) • Unifix cubes (all colors) • Worksheets (see black-line master 5)

The children fill a margarine tub with cubes. They then sort the cubes by color and determine the number of cubes of each color. They write the numerals on worksheet 5. If they do not have any cubes of a particular color, they write 0.

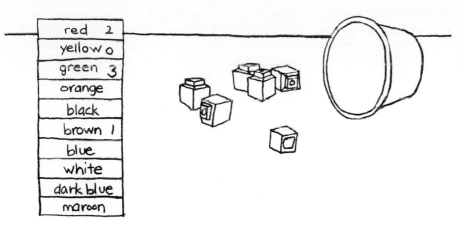

red	2
yellow	o
green	3
orange	
black	
brown	1
blue	
white	
dark blue	
maroon	

(Have a chart with a colored dot by each word to which children can refer.)

ANALYZING AND ASSESSING YOUR CHILDREN'S NEEDS

The most accurate and in-depth assessment of your children will come from your observations while they are working and learning. Concepts such as conservation of number, relationships between numbers, and the building of visual images of quantities develop over months and years and cannot be pre- and post-tested in a nine-week instructional period. We need to be in tune with this growth in our students so that we can provide the kinds of experiences that will help them develop these understandings.

Some skills *can* be pre- and post-tested, and you can expect some degree of mastery of them. The following skills can be quite easily assessed and are valuable pieces of information that you should have about each of your children.

Rote sequence: How high can they count orally?

One-to-one correspondence: Do they count each object once and only once consistently?

Association of numeral and quantity: Can they say the names of the numerals and build the appropriate sets when shown the numerals?

Writing numerals: Can they write the numerals with ease to name a particular set?

Once you have determined which children need help with the above skills, you will find the activities in this chapter will aid in the attainment of these skills. However, if these are the only goals we have set for our children, then we are shortchanging them. The activities in this chapter have been selected so that many of them will promote the development not only of basic counting and numeral recognition skills but also will offer the opportunity for children to develop conservation, comparison of number, and, in general, a stronger sense of number.

The activities in this chapter are designed to be played at a variety of levels. The level of many can be varied simply by changing the size of the number being worked with. Other tasks will be approached by children of different levels of understanding in a variety of ways. For example, in the game "Grow and Shrink," in which the children build on working space paper the numbers that the teacher says, the following kinds of behavior may occur which can give the teacher clues to the children's thinking.

After the teacher says "Eight," Bill counts out the eight cubes carefully, with great concentration, and arranges them on his working space paper. Linda and Paul put out the eight cubes easily. Linda knows there are five dots in the first row on her working space paper. She puts out five and then counts on ("six, seven, eight") to complete the task. Paul puts out eight very quickly. He knows there are five dots in the first row and that you need three more to make eight. Next, the teacher says "Six." Bill removes the eight cubes he has on his board and begins to count out six. Linda knows she needs the five in the first row of dots and that one more would make six. She takes off the extra cubes without particularly noticing how many extras she had. Paul takes off two cubes. He knows that six is two less than eight.

Both Bill and Linda are benefiting from this activity. For Bill it is a counting game, and he still needs that practice. In time he will begin to approach the task in a way that is similar to the way Linda is approaching it now. For Linda it is an opportunity to focus on the relationships between numbers and internalize them. It may or may not be appropriate for Paul. The teacher would have to note Paul's level of interest and involvement. If he appears bored and uninterested, then he probably needs a new challenge. If he appears involved and pleased with the task, then he probably is at a stage where the practice is still beneficial. It's like beginning readers who read stories that are "easy for them." It can be pure pleasure and a confirmation of what they have learned.

It is clear that we will not meet individual children's needs simply by creating a list of skills and checking them off as the child masters them. Developing concepts is much more complicated, and we must continually observe our children for clues that will tell us what they are thinking.

COUNTING

Emily knows the rote counting sequence to twenty, but when asked to count cubes she says the words and points randomly at the cubes. She has no idea yet that one word goes with one object.

Chris is asked to count some cubes and begins saying "1, 2, 3, 4" before he even starts pointing to the first object. He doesn't say the correct counting sequence past six. He doesn't know what counting is about.

Nancy is asked to predict how many there are in a group of cubes.

Although there are eight cubes, she has predicted five so counted in such a way (skipping cubes) to get to five so she could be "right." Nancy has not yet developed a strong notion of what counting is about.

John counts correctly, but when asked "How many?" he doesn't know. He seems to be going through a ritual of pointing and saying words but apparently does not relate this ritual firmly with an idea of number. The last word he says in a sequence when he is "counting" is just a word—so easily forgotten.

Andy knows the counting sequence, but when told "Hand me six cubes" he keeps counting past six. Six as a quantity has no meaning for him as yet.

Jim is asked to count cubes. He says, "One, two, three, five, nine, eight, seven."

The above children need lots of practice counting. They should be grouped together for small-group, teacher-directed lessons to play the games in Section I: Developing the Concept. They should focus in the beginning on groups of four or less and move to counting larger groups as they gain skill and confidence with groups of two, three, and four. In order to assign independent activities, the teacher should find out how high each of the children can count successfully. Some who can work with groups of four or even more with teacher direction may need to count out groups of two when working independently. The counting boards can provide the opportunity for repetitious counting in a way that allows the children to experience some variety and use of imagination. The teacher will find that children will move to larger numbers more easily if they are able to focus their attention on just a few numbers for a while and are not overwhelmed with the task of counting to ten.

NUMBER SENSE

Both Diane and Susan count any group of objects two or three times, just to be sure they counted right. Sometimes they stop in the middle of counting and start over. Counting is more than a ritual of saying words for them now. They know what to do and want to be sure they do it right and get the correct number.

No matter how small the group of objects (even two or three), Robert counts from one each time. It's as if he thinks counting is always pointing at every object. He either has not developed instant recognition of small groups or has not realized it's okay not to count (that is, point) if he already knows how many.

Judy counts correctly, carefully pointing at each object; however, when asked to count a different group of cubes, she no longer points at each object as she counts. Children will often be inconsistent when learning something new. Judy's attention may have been on saying the sequence correctly and not on one-to-one correspondence. This happens more frequently when she is counting "big" numbers.

Mark was making trains of five. Most of the trains he made indeed were five cubes long, but one of them was only four cubes long. He did not seem to notice or be concerned about the difference in length because the trains were not lying close to each other. When he was recounting all the trains and came upon the one with four cubes, he was very surprised to find it. After all, he had made trains of five. To fix it, he broke one off and then was even more surprised to find there were only three cubes. He smiled and knew immediately what to do. He put the "four" cube back on and added one more to make five. Mark is just beginning to discover how the numbers work.

The children described above have the beginnings of understanding what number is. They know how to count and can answer you when asked "How many?" but they need more practice and experience to strengthen these developing concepts. The teacher should choose a variety of games from Section I and watch the children's understanding of number grow.

◆ ◆ ◆

Randy was making trains. His were to be six cubes long. He had already started breaking the trains apart when the teacher asked him to count his trains for her. He counted the train he had already broken into two parts and was amazed to find there were only five cubes. He put them back together, thinking that must be the problem, and was surprised to find he still had only five cubes.

Laurie counted eight objects correctly. When one was removed, she recounted and got eight again and seemed perfectly content with that, not noticing the inconsistency.

Linda was making trains of eight. She was counting accurately and confidently, and all the trains she made were indeed eight cubes in length. She was asked to break one of the trains into a margarine tub and tell how many were in the tub when she finished. She said, "I don't know how many there are now." Breaking the cubes apart made them look different, and she was no longer sure how many there were. Joey did the same task. He was willing to make a guess. He thought maybe there was six now, or maybe seven. When Todd did the task,

he said eight immediately and looked as though he wondered why the teacher bothered to ask when the answer was so obvious. At another time, however, when he did a task with seventeen cubes, he was no longer sure of how many cubes there were.

For the children described above, the concept of conservation is still elusive. They will develop this sense over time as long as they are able to deal with real objects in a variety of situations. What seems so illogical to us is a natural part of the child's developing sense of number. It is not necessary to protect children from further number experiences until they master conservation, because it is through a combination of experiences that understanding can develop. Learning about numbers does not occur in nice sequential steps requiring mastery of one before moving to another. What specific activities these children need depends on the level of skill development they have reached in other areas. It is essential that they have lots of opportunities to work in a variety of situations with real objects.

WORKING WITH NUMERALS

Brad knows the names of the numerals so has been given a set of numeral cards, and he is attempting to count out the appropriate number of cubes. Many of his counting boards have the wrong number of cubes on them. When the teacher asks him to count the cubes, she finds Brad can say the rote counting sequence very accurately. But rather than counting out the cubes carefully, he simply says the words while he is putting cubes on the boards. He doesn't seem to have the idea of one-to-one correspondence. Brad probably has been praised for saying the counting sequence and for saying the names of the numerals. What no one had yet noticed is that he doesn't know what the words and symbols stand for.

Children like Brad who are having difficulty counting should not be dealing with symbols. Symbols should be used to label what is already known conceptually. Brad needs to work in a small, teacher-directed group with others who need to focus on numbers to four or five. Almost all the activities in the first section can be played with numbers to five. Even though he may be more insecure than others in the group, Brad will have the support of the teacher and other children who will be modeling what to do. Eventually, he will be able to do the counting tasks on his own. During those times when Brad needs to work independently, he should practice counting only to those numbers with which he can be successful. He may need to start by counting out two cubes onto sets of counting boards. He can have a card with his name on it and the number of dots indicating what he should count to. When counting to two is easy for him, the teacher can add a dot to his card, and he can practice counting to three. When Brad can count to four or five, he will be ready to deal with the other independent activities in Section I.

◆ ◆ ◆

Todd has been counting out cubes onto the counting boards, and he has made some mistakes. The teacher asks him to count some of the sets but is careful not to ask Todd just to count the boards with the mistakes. When he recounts one that is wrong, he recognizes it immediately and fixes it before the teacher can say a word.

Susan has mistakes on her counting boards, too. The teacher asks her to recount, making sure he is not asking her to recount only the groups with mistakes. When Susan counts for the teacher, she counts all the little dots on the card before she says the numeral. It appears she does not know many of the numerals yet.

Joel is playing "Build a City" with a friend. When it is his turn to roll the die, he looks to his friend to tell him the numeral. The teacher remembers that yesterday, when Joel was working with the counting boards, he was counting the dots on the numeral cards for each board.

When he asks Joel to tell him the numerals, he doesn't seem sure of any except one. All of the other children in the group that Joel has been working with have learned the numerals, but Joel has not. Joel counts accurately and does well playing the games in which no symbols are involved. He is having unexpected difficulty with the symbols.

Dealing with numerals can pose a variety of problems for children. For some who are just learning, their concentration on the numerals causes them to make counting mistakes they wouldn't have made if they had not been dealing with the symbols (a new task for them). Time and practice will take care of that. Some children may be very insecure working with numerals. In some cases, as in Susan's, limiting the number of new symbols she is to work with and adding new symbols gradually may help. Occasionally, children who are working well conceptually have extreme difficulty working with symbols. Such children should not be prevented from experiencing more complex concepts just because the symbols are difficult. Special help can be offered such as sandpaper numerals and three-dimensional magnetic numerals that allow children to feel the shapes of the symbols. In the meantime, children who are ready to explore new ideas conceptually should not be held back. Provide whatever aids are necessary for the children to complete assigned tasks (number lines, dotted numeral cards, etc.).

PROVIDING CHALLENGES

Laura is counting out cubes onto counting boards. She looks up eagerly when the teacher stops by to see her and begins to tell stories about each of her boards. She is pretending to count children onto a playground. Laura has put the numerals she has been using in order. She has three 6's. "Look," she says, "I put four children on the swings and two in the sandbox for this 6. In this playground three kids are swinging and three are on the slide, and on this board only one is swinging and the rest are resting." Laura is finding different ways to arrange the cubes so the activity is more interesting for her. She seems to be enjoying the fantasy world she has created.

Usually, one or two children in a class seem to learn faster and are more advanced conceptually than the other children. It is a challenge to meet their needs. For most children, it is not good for them to be isolated from the group and working alone. Yet there seems to be no point in having them do tasks that are too easy for them. The solution is to provide tasks that require problem solving and creativity or that can be experienced at many levels. The counting boards provide that kind of opportunity for Laura. She is being creative in the stories she is making up for each board. Mathematically, she is exploring the combinations of six that will be the foundation for internalizing basic facts. The teacher can pose questions to Laura that will keep her challenged, such as: "How many different ways can you arrange six cubes? eight cubes? three cubes? Can you arrange the cubes so I can tell quickly how many there are when I walk by?" The Unifix puzzles also have the potential for problem solving. For example, you could tell each child to take three puzzles and put them in order by the number of cubes they think each puzzle will need. Have them check their predictions. Have the children sort the puzzles into "easy to tell how many by looking" and "hard to tell by looking."

Children often make up their own challenges such as Laura was doing on her own with the counting boards. They will be seeing relationships and noticing qualities of numbers on their own simply because of the nature of the activities that use materials such as the cubes. In many other seemingly simple counting activities, number relationships are there to be discovered. You do not need to spend a great deal of time planning special activities for special children, but you should provide opportunities for them to modify the activities in their own ways. Each child's level of involvement or lack of it will be the clue to deciding if you need to provide them with a challenge.

◆ ◆ ◆

No matter how hard you try, you can't know with absolute certainty exactly what the next step of learning should be for each child you are expected to teach. Therefore, it is essential that you provide a range of meaningful experiences that can be approached in a variety of ways and will serve to stretch all your children's thinking. You need to provide opportunities for children to choose activities for themselves. And you must continually look to the children for the clues that will help determine whether to offer tasks that are more simple or more complex.

CHAPTER TWO

If your textbook or workbook objectives
are:
- What number comes after
- What number comes before
- Counting by twos, fives, tens, etc.
- What comes next in the sequence

Then you are dealing with:

Pattern

Section I. Developing the
Concept of Pattern

Section II. Extending the
Concept of Pattern

WHAT YOU NEED TO KNOW ABOUT PATTERN

Much of our lives is spent in an active search to make sense of things—to organize and sort things out. When, as children, we are able to get a sense of the basic order of things, we are able to predict—to count on things happening—and thus to become more secure and more confident. We learn that morning always follows night, that second grade comes after first, that a flash of lightning will be followed by a roll of thunder.

Seeing patterns in the way things work is an incredibly powerful learning tool that most of us have developed intuitively to some degree. If we add *s* to words to make them plural, if we know a dog we have never seen before is a dog and not a cat, if we know that red lights always turn green and that green lights turn yellow, we are using our sense of pattern.

It may surprise some of you to learn that mathematicians say that mathematics is the study of pattern. They say that pattern is the basis on which our number system was created. Most of us missed that truth, because for us mathematics has been a series of rules and steps to follow so we could get answers to teachers' questions. We did not try to look for the underlying order or sense of things in mathematics. If you have not yet discovered the pattern in number, learning mathematics has been much more difficult than it needed to be, and you have missed much of the beauty that is in mathematics.

You can give your students a sense of the beauty and order that is mathematics. You can give them the confidence that comes when things are predictable. The activities in this chapter are designed to give children such opportunities.

Young children need to experience patterns first in motion, color, design, and arrangement. Then from these experiences comes the discovery of the pattern in number.

This chapter has been divided into two sections. The first section presents activities for developing the concept of pattern. These activities can be used to introduce pattern to children in any primary grade, because the complexity of the patterns can be varied according to the needs of the children. The second section presents activities for extending the concept of pattern. The activities in this section involve somewhat more complicated patterns than those in the first section.

Section I: Developing the Concept of Pattern

The concept of pattern should be experienced as an ongoing part of your math program, weaving in and out of the work you are doing with other math concepts. There may be periods of time when you will concentrate on pattern activities for a week or two, but be sure you continue to include pattern activities on occasion throughout the year.

Don't hesitate to repeat some of the introductory activities as well as introduce some of the extending pattern activities. You will see that the patterns the children create during the year will become more and more complex. The same activity done in November will change in character when repeated in March, because the children will be able to work at a higher level.

Teacher-Directed Activities

Introduce your children to the concept of pattern through rhythmic patterns that take just three to four minutes. They can be done at any time of day for several days or weeks before you begin using Unifix cubes to work with the concept of pattern. Later on, you will want to set aside several days for concentrated pattern work with Unifix cubes. After this period of time, during which the children have had the opportunity to experience a variety of pattern activities, continue to present pattern activities periodically for the remainder of the year.

RHYTHMIC PATTERNS*

Materials: None required.

Begin a rhythmic pattern, and ask the children to join you after a few motions.

For example: Clap, clap, slap (legs); Clap, clap, slap; Clap, clap, slap . . .

Continue the pattern for at least thirty to sixty seconds. There is something very satisfying about the repetition of the rhythm. A long repetition will help some of the children who are having difficulty begin to feel the pattern inside, even if they cannot yet do it perfectly with their bodies.

*Based on MATHEMATICS THEIR WAY, p. 21

By saying the names of the motions or body parts as you do the pattern.

Do: Clap, clap, slap; Clap, clap, slap; . . .
Say: Clap, clap, slap; Clap, clap, slap; . . .

By adding a varied motion to one part of the pattern.

This time let's nod our head when we slap.

By labeling the pattern with ABC's to analyze and describe the pattern (see note below).

Do: Clap, clap, slap; Clap, clap, slap;
Say: A, A, B; A, A, B

Spend three or four minutes a day working with rhythmic patterns. Each day choose another pattern to do with the children. Do some simple patterns and some more complex patterns. Do not expect or wait for all the children to master a pattern before moving on to others. The children will improve over time if given enough variety.

Do not single out individual children for special help. A child who is having difficulty with a pattern may be less inclined to try if the unsuccessful attempts are given attention. Provide for individual needs by interspersing simple patterns in with the more complicated ones. When children have difficulty, teachers are often tempted to slow the pattern down. This actually makes the pattern more difficult to "feel." Keep the pattern moving at about the rate of children's heartbeats.

Examples of other patterns you can do on succeeding days:

Clap, snap (fingers); clap, snap; clap, snap; etc.
Stamp, stamp, stamp, stamp, stamp, stamp, etc.
Nod, nod, clap; nod, nod, clap; etc.
Nose, shoulders, shoulders; nose, shoulders, shoulders; etc.
    (Touch the body parts as you say them.)
Slap, clap, shoulders; slap, clap, shoulders; slap, clap, shoulders, etc.
Nose, nose, nose, jump; nose, nose, nose, jump; etc.
Slap (legs), slap, clap, clap; slap, slap, clap, clap; etc.
Chin, chin, chin, chin, chin, etc.

Using the same pattern, begin varying it slightly:

By verbalizing one (not all) of the motions.

Do: Clap, clap, slap; Clap, clap. slap; . . .
Say: _____ slap; _____ slap; . .

Note: Naming the parts of the pattern orally using ABC's is a very helpful tool for children. Many find it easier to do some patterns if they label the motions in the following manner.

For example:

Shoulders, nose; shoulders, nose; etc.
   A   B   A   B

Snap, snap, clap, clap; snap, snap, clap, clap; etc.
 A   A   B   B   A   A   B   B

Nose, nose, clap, jump; nose, nose, clap, jump; etc.
 A   A   B   C   A   A   B   C

The labels also serve as a useful tool for naming and comparing patterns. For example, all of the following are AB patterns:

Snap, clap; snap, clap; etc.
Hair, chin; hair, chin; hair, chin; etc.
Arm up, arm down; up, down; up, down; etc.

The following patterns are AABB patterns:

Flap (arms), flap, clap, clap; etc.
Slap, slap, stamp, stamp; slap, slap, stamp, stamp; etc.
Tall, tall (stretch up), short, short (squat down); etc.

The usefulness of the ABC labels will become more apparent as you work with the other pattern activities described in this chapter.

INTERPRETING RHYTHMIC PATTERNS WITH UNIFIX CUBES*

After several days of doing rhythmic patterns, introduce this activity. Have the children seated in a circle on the floor (or around a table) so they can see each other. Dump out the cubes so all the children have access to many different colors. Begin a rhythmic pattern and have the children join in. After a minute or so, stop the motion and tell the children to make that pattern by snapping the cubes together.

For example: *Join in as soon as you know the pattern.*

Clap, slap, slap; clap, slap, slap; etc.

Now, see if you can make that pattern by snapping the cubes together. Make your pattern as long as you can.

Observe the children as they work. Do not expect everyone to understand your directions, and do not try to explain what to do. Allow those who don't understand time to observe others and figure it out for themselves.

Next, find a pattern that has been done correctly, and direct the children's attention to that pattern.

For example:

Let's look at Linda's pattern. What colors did she use? Linda, point at the cubes while we say the colors.

Yellow, blue, blue; yellow, blue, blue, etc.

(To help the children get the idea that the pattern can go on and on, continue to say the colors past the last cube.)

To help them see that there are lots of different ways to interpret the *clap, slap, slap* pattern, find another pattern that has been done correctly and point it out to the children.

For example:

Barbara used different colors to make our clap, slap, slap pattern. Barbara, point to the cubes in your pattern while we say the colors.

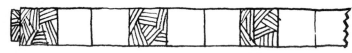

Green, red, red; green, red, red; green, red, red; etc.

Now, let's say the pattern using ABC's.

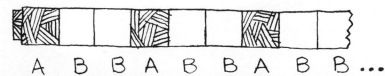

A B B A B B A B B ...

At this point, many children who did not understand before will have figured out what you were asking them to do. Do not expect all the children to understand; it will take several experiences for everyone to understand how to interpret a variety of patterns with the cubes. Expect some confusion along the way as the children sort it out.

On another day when you want to interpret patterns again, be sure to use a different pattern (not ABB) for the next experience or some children will think they are supposed to always make ABB patterns with the cubes. Try an ABC pattern or an AABB pattern next.

Extensions: a. Have the children use their own ideas to make up patterns. Some of the children can then share their patterns with the others.

For example:

Today everyone gets to make any pattern they can think of.

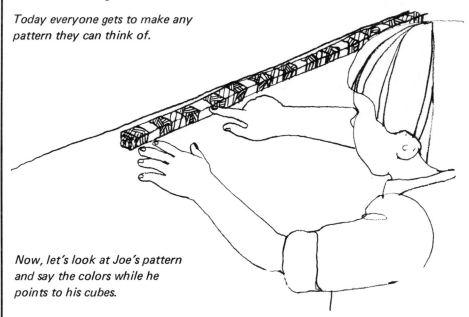

Now, let's look at Joe's pattern and say the colors while he points to his cubes.

Red, white, red, yellow; red, white, red, yellow . . .

What would Joe's pattern be if we said it with ABC's?

ABAC; ABAC; ABAC; ABAC; etc.

Let's look at Martha's pattern.
Let's clap every time we say
red.

White, white, red, green; white, white,
red, green; white, white, etc.

b. Have the children work by themselves to find all the different patterns they can. As they discover new ways to make patterns using their Unifix cubes, have them show you their discoveries, and record the patterns on a chart. The chart can be added to on succeeding days.

For example:

Patterns

ABC - Peter
ABAC - Wes
AAAB - Jackie

I have a different pattern. It's
AABCD. It can go on the chart.

c. Give the children a particular pattern such as AABB and have them work individually to see how many different ways they can make that pattern. Record the ways on a chart.

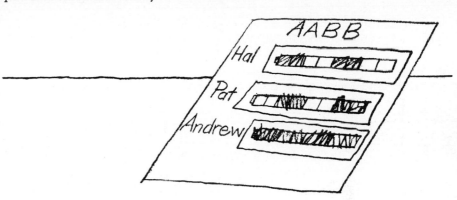

AABB
Hal
Pat
Andrew

PATTERN ARRANGEMENTS

Materials: Unifix cubes (sorted by color)

Have the children explore ways to make patterns without snapping the cubes together. Allow each child to use only one color. This will force them to think about ways to make patterns that do not rely on color. After they have had time to make patterns, have them discuss with the others what they did. This is a wonderful opportunity to develop language along with pattern.

For example:

We could put them up and down like this and make a pattern. That's up, down, up, down; up, down, etc.

I know another way to describe it; top, bottom, top, bottom.

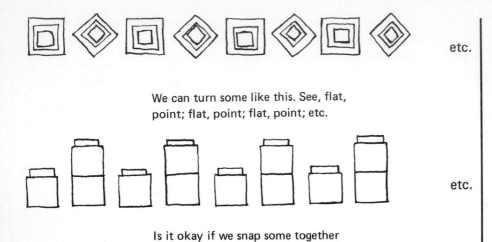

We can turn some like this. See, flat, point; flat, point; flat, point; etc.

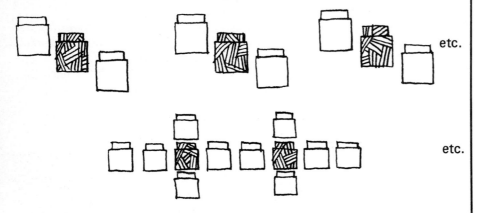

Is it okay if we snap some together but not all, like this?

After they have discovered the possibilities using one color, allow them to use more than one color if they wish.

etc.

etc.

Independent Activities

Children need opportunities to build and create patterns on their own. You will see some children making very simple patterns while others make more complicated ones. Children of all levels can work side by side and learn from each other. You can add variety to the independent activity time by allowing children sometimes to create their own patterns with no direction from you, sometimes to use task cards and copy and extend those patterns, and sometimes to create patterns and then record their creations in various ways. Encourage the children to

make long patterns, because it is through repetition that they begin to get a true sense that patterns go on and on.

PATTERN TRAIN TASK CARDS

Materials: Unifix cubes • Pattern train task cards (using black-line master 32 run off pattern train outlines, then color in a variety of patterns and mount them on tagboard and laminate.

Have the children choose a card, copy the pattern with the Unifix cubes and extend that pattern.

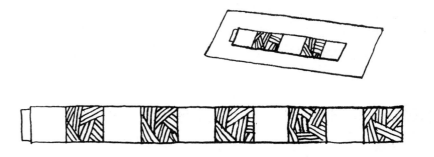

Variation: The children can make task cards for the other children to copy with the Unifix cubes and extend. Mount the patterns the children have colored in on worksheets (black-line master 32) on tagboard and preserve them by laminating or covering with clear contact paper.

PATTERN ARRANGEMENT TASK CARDS

Materials: Unifix cubes · Task cards (teacher-made—see example below)

Provide pattern arrangement cards for the children to copy with the Unifix cubes and then extend.

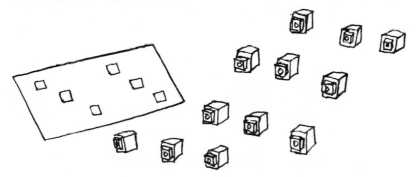

Variation: Have the children create task cards of their ideas for the other children to use when working independently. They can make the cards by:

a. pasting squares of paper that match their cubes onto 3 x 9 pieces of tag

b. drawing the patterns using a square template (see p. 215 for directions for making)

c. using the commercially available Unifix Album of Gummed Colored Strips (see p. 217)

RHYTHMIC PATTERN CARDS*

Materials: Unifix cubes • Task cards (run off black-line master 30) Cut out various motions pictured on the master and arrange them in a variety of patterns

Have the children imitate the motions pictured on the task cards and then translate the pattern to Unifix cubes.

For example:

Variation: The children can make up a simplified version of the rhythmic pattern cards on long strips of paper (butcher paper, shelf paper, or computer printout paper) by tracing around their feet to represent stamping and around their hands to represent clapping. These can be hung in the room, and the other children can imitate the motions and make the patterns with cubes.

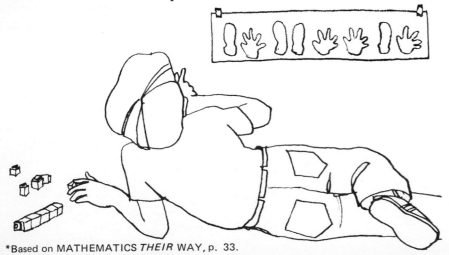

*Based on MATHEMATICS *THEIR* WAY, p. 33.

ABC PATTERN CARDS

Materials: Unifix cubes • ABC pattern cards (teacher-made—see examples)

Provide ABC pattern cards for the children to interpret with Unifix cubes. It will be interesting for the children to find out how many different ways the same pattern can be interpreted with the cubes.

For example:

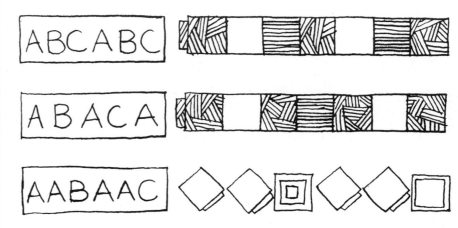

RECORDING THE PATTERNS

Materials: Unifix cubes • Worksheets (see black-line master 32)

The children can record the patterns they make by coloring in the outlines on the worksheets. You can extend these activities by having the children label these patterns with ABC's.

Section II: Extending the Concept of Pattern

Introduce the pattern activities in this section after your children have had success with the activities in the first section. These activities can be presented whenever you wish to provide variety and a change of pace while increasing the children's understanding of mathematics.

INCREASING PATTERNS*

The following activities allow children to use their sense of pattern to build designs that increase in an orderly, predictable way. The nature of the designs is such that number relationships are also highlighted.

Introducing Increasing Patterns

Materials: Unifix cubes (sorted by color)

Call together a small group of children. Have them sit directly across from you so they will all get the same view of the patterns you build. Build a design with the Unifix cubes.

For example:

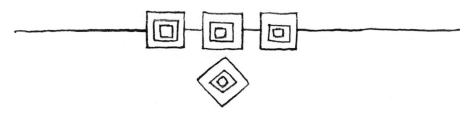

Tell the children you are going to make the design grow step by step. Point to each part of your design, and ask the children to tell you how many cubes are in each part.

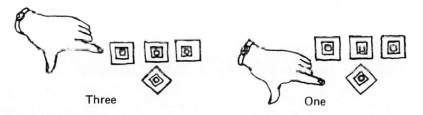

Three   One

*Based on MATHEMATICS *THEIR* WAY, p. 261.

Say, "This is my first design. Watch to see what I do to make the next design." (Keep the first design intact. Make the second design by adding to each part one more cube than the original design has.)

Point to each part of the second design, and have the children tell you how many are in each part.

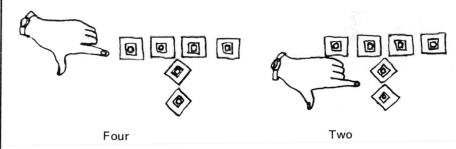

Four   Two

Tell the children to watch you again and see what you do to make the next design. (The third design has one cube more in each part than the second design.)

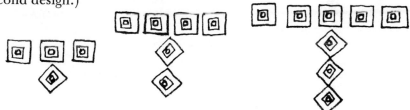

Point to each part of the design, and have the children tell you the number of cubes in each part.

Five   Three

Ask the children to build what they think will be the next step. After they have had time to build their ideas, you also build the next step so they can check their work.

Continue in the same way for several steps, each time asking the children to build what they think the next design will be.

After you have made six to eight designs, point to each part of each of the designs, starting with the first one, and have the children tell how many there are.

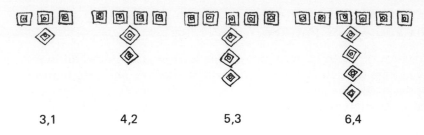

3,1   4,2   5,3   6,4

Ask the children to predict the pattern beyond the cubes they can see. Ask, "What do you think will come next, and next, and next. . . . "

7, 5 8, 6 9, 7 10, 8 11, 9 12, 10 etc.

Continue on succeeding days to build many other patterns.

Examples:

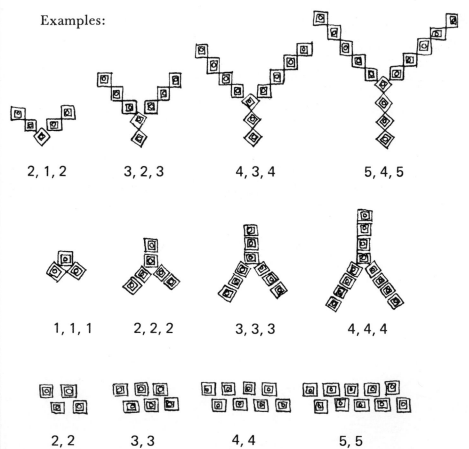

2, 1, 2   3, 2, 3   4, 3, 4   5, 4, 5

1, 1, 1   2, 2, 2   3, 3, 3   4, 4, 4

2, 2   3, 3   4, 4   5, 5

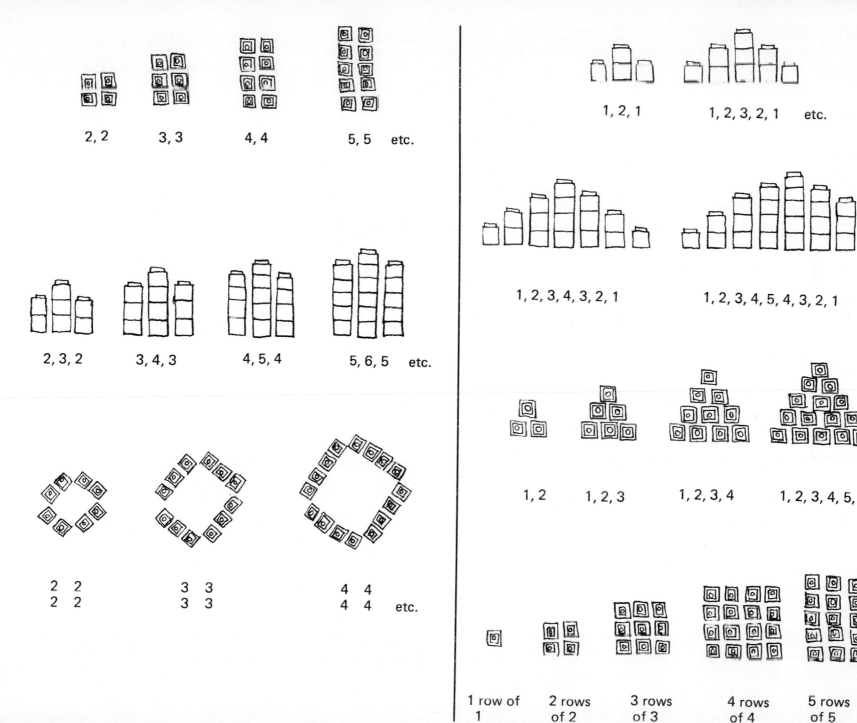

2, 2   3, 3   4, 4   5, 5   etc.

2, 3, 2   3, 4, 3   4, 5, 4   5, 6, 5   etc.

1, 2, 1   1, 2, 3, 2, 1   etc.

1, 2, 3, 4, 3, 2, 1   1, 2, 3, 4, 5, 4, 3, 2, 1   etc.

2 2   3 3   4 4
2 2   3 3   4 4   etc.

1, 2   1, 2, 3   1, 2, 3, 4   1, 2, 3, 4, 5,   etc.

1 row of   2 rows   3 rows   4 rows   5 rows
1   of 2   of 3   of 4   of 5   etc.

Note: The same design can grow in different ways. Any way is right, as long as you are consistent after you have chosen the way you want to increase it.

For example:

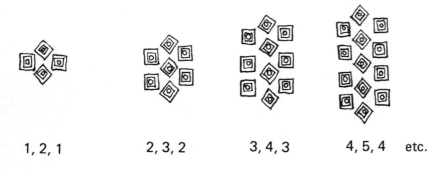

1, 2, 1   2, 3, 2   3, 4, 3   4, 5, 4   etc.

or

1 1 1 1   2 2 2 2

3 3 3 3   4 4 4 4 etc.

Working Independently with Increasing Patterns

After the children have had many experiences building patterns during small-group time, allow them the opportunity to create patterns independently.

Increasing Pattern Task Cards

Materials: Unifix cubes · Task cards (teacher-made—see examples)

There are three types of task cards that children can copy and extend.

Three-Step Cards

When you want the children to build the pattern in a certain way, you need to include at least three steps so they can tell how the pattern is to grow.

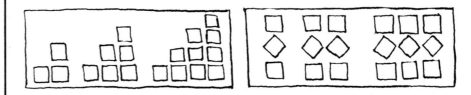

One-Step Cards

Provide task cards that give the first step of the pattern. Let the children increase the pattern any way they like, as long as they are consistent.

For example:

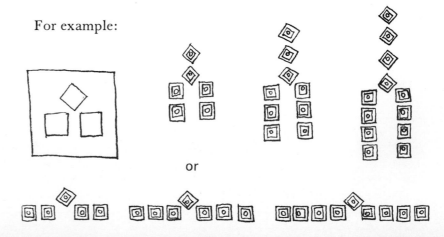

or

Interpreting Numbers Task Cards

Provide task cards that have the number of cubes that belong in each step. The children can interpret the numbers in a variety of ways.

For example:

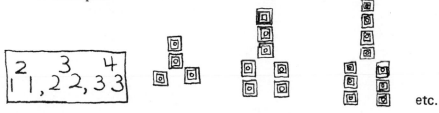

or

or

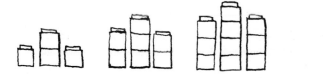

etc.

Creating Their Own Patterns

Materials: Unifix cubes

Allow the children the opportunity to make up their own designs and have them grow in any way as long as they are consistent.

Recording the Patterns

Materials: Unifix cubes · Various recording materials (see below)

Children can record their designs on shelf paper, butcher paper, computer printout paper, or construction paper, which they tape together. They can make the squares using a square template (see p. 215 for directions for making) or stickers commercially available in the Unifix Album of Gummed Colored Strips (see p. 217).

Looking for Number Patterns in the Increasing Patterns' Designs

Materials: None required

Children have to look at the parts of the increasing patterns' designs in order to make the designs grow in a consistent way. But there are many more patterns to be discovered. If they reexamine their patterns in new ways, they will be able to discover a variety of interesting number patterns.

In the following designs, the children can look for patterns occurring in the "arms," middle, and total number of cubes used in each step:

The pattern for the "arms":

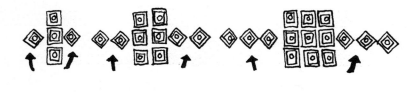

two   four   six

eight   etc.

The pattern for the "middle":

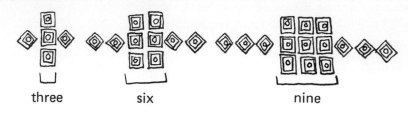

three   six   nine

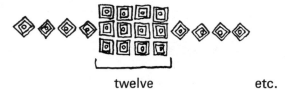

twelve   etc.

The pattern for the "totals":

five   ten

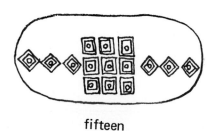

fifteen

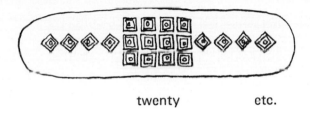

twenty   etc.

Looking for Patterns Formed by Parts of the Design

Materials: Unifix cubes • 00–99 chart (see black-line master 70)

The children will be able to predict the number patterns far beyond what they can see by looking at the cubes they have actually used in their designs if they record the number patterns on a 00–99 chart (black-line master 70) or a number line. The recording of the patterns in this way will have greater impact on most children if they have had experiences with the place-value activities in Chapter Five.

Have a group of children work with you to create six to eight steps of an increasing pattern. Then have them look for patterns that are formed by parts of the designs.

For example: Let's count the cubes that are in the top row of our designs.

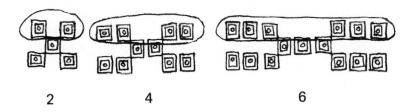

2   4   6

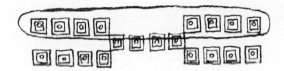

8

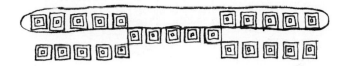

10

As the children report the numbers, circle them on a 00–99 chart.

00	01	02	03	04	05	06	07	08	09
10	11	12	13	14	15	16	17	18	19
20	21	22	23	24	25	26	27	28	29
30	31	32	33	34	35	36	37	38	39
40	41	42	43	44	45	46	47	48	49
50	51	52	53	54	55	56	57	58	59
60	61	62	63	64	65	66	67	68	69
70	71	72	73	74	75	76	77	78	79

(02, 04, 06, 08, and 10 are circled on the chart.)

After they have reported all the numbers they can by counting the cubes in the designs, have them predict the numbers that would result if they continued to make more designs. Circle those numbers also on the 00–99 chart. (The power of pattern is in the fact that you can make predictions beyond the physical evidence you have present.)

(00)	01	(02)	03	(04)	05	(06)	07	(08)	09
(10)	11	(12)	13	(14)	15	(16)	17	(18)	19
(20)	21	(22)	23	(24)	25	(26)	27	(28)	29
(30)	31	(32)	33	(34)	35	(36)	37	(38)	39
(40)	41	(42)	43	(44)	45	(46)	47	(48)	49
(50)	51	(52)	53	(54)	55	(56)	57	(58)	59
(60)	61	(62)	63	(64)	65	(66)	67	(68)	69
70	7				75	(76)	77	7	

Note: You can help the children predict the numbers in a pattern beyond what they build with the cubes by showing them the following technique.

Tell the children to read the pattern they have already made by saying *yes* for each number circled and *no* for each number not circled. For example: Starting with the first number that is circled, have them read "Yes, no, yes, no; yes, no, yes, no; yes, no, yes, no; etc." Tell them to keep saying the yes, no, yes, no pattern and circling the numbers they are pointing to each time they say *yes*.

Looking for Patterns Formed by the Total Cubes in Each Step of the Design

Materials: Unifix cubes • 00–99 chart

Have the children look at the design again—this time to see how many cubes were used to make each step of the design. Circle those numbers on another 00–99 chart. Again have them predict the numbers beyond the physical evidence presented by the designs themselves.

5   10   15

20

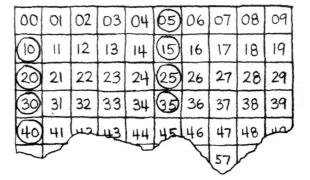

00	01	02	03	04	(05)	06	07	08	09
(10)	11	12	13	14	(15)	16	17	18	19
(20)	21	22	23	24	(25)	26	27	28	29
(30)	31	32	33	34	(35)	36	37	38	39
(40)	41	42	43	44	45	46	47	48	49
							57		

Labeling the Number Patterns with Color

Materials: Unifix cubes • 00–99 chart • Various strips of colored construction paper (3″ × 18″) • Crayons to match the paper • Clear tape

As the children work with increasing designs, they will be discovering a variety of patterns and then rediscovering those same patterns in other designs. To help them recognize the same patterns occurring and reoccurring, have the class choose colors to label each of the patterns they encounter. For example, they might choose the color blue to label the 2, 4, 6, 8 pattern. The numbers in that pattern can be written on strips of blue construction paper and hung in the room for reference. They might choose the color red for the 5, 10, 15, 20 pattern. This pattern can be written on strips of red construction paper and hung in the room. As new patterns are discovered, new colors are chosen. Each time they find a pattern, they check to see if it is a pattern they have already worked with or a new pattern.

For example: Various strips of construction paper from which the children can choose, along with crayons to match each of the colors have been made available. The children have developed the following increasing design pattern.

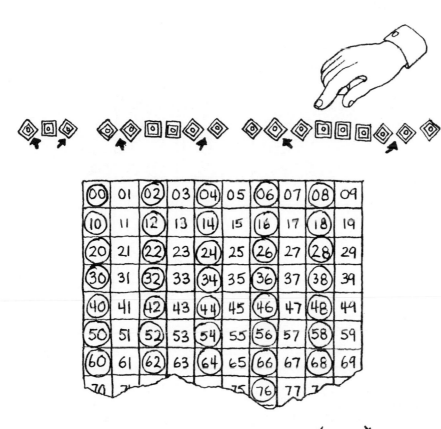

They have discovered the pattern for the "arms" is a 2, 4, 6, 8 pattern. They have circled the pattern on a 00–99 chart and have predicted what the pattern would be beyond the cubes they have used.

We want to choose a color for our 2, 4, 6, 8 pattern. Every time we see this pattern, we will call it that color. Who has an idea?

Blue.

Okay, the 2, 4, 6, 8 pattern will be our blue pattern. Read the numbers we circled on our 00–99 chart, and I will write them on these blue strips of paper.

Before you begin writing, look at the 00–99 chart and see if it looks like zero should be included in the pattern. If so, begin with zero when writing the numbers.

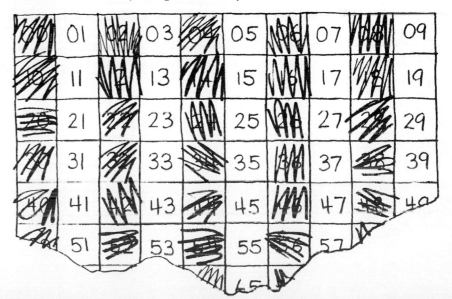

Tape strips together.

The children should then color in the pattern on a 00–99 chart (see black-line master 70) using a blue crayon.

As the children discover other patterns, choose other colors and record those patterns on construction paper strips.

For example: The following designs have been built.

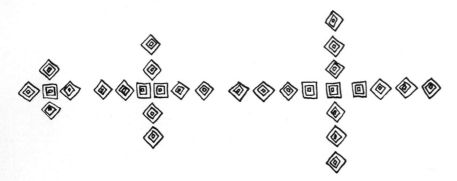

The total cubes used in each design make a 5, 10, 15 pattern. The children picked red to be the 5, 10, 15 pattern. The numbers are written on strips of red construction paper and are hung in the room for reference.

On succeeding days, the children will make more designs and look for patterns. If they discover a new pattern, they will choose a new color and new strips will be made and hung up. If they discover a design has the same number pattern as on one of the strips already up, they will color in a 00–99 chart, using the appropriate color.

For example:

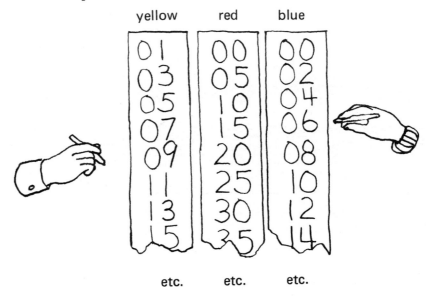

My pattern is a 3, 6, 9 pattern. We need to choose a color for this pattern.

I made a 5, 10, 15 pattern. That's the red one. I need to color my 00–99 chart with red.

UNIFIX BREAK-APARTS*

Materials: Unifix cubes

Many interesting and beautiful patterns can be created by breaking long Unifix train patterns into various shorter lengths.

The first step is to have a group of children work together to create a very long train pattern.

For example: Today we are all going to make some patterns. We are going to work together so we can make it very long. The pattern is yellow, yellow, green. Make it like this so the green is on top when you stand it up.

Each child makes several little trains. Then all the little trains are snapped together to make one very long train.

After the pattern has been made, have the children sit facing you. Spread out tiny number cards from two to eight and ask the children what number they would like to break the train into.

Four.

*Based on MATHEMATICS *THEIR* WAY, p. 267.

Count with me, and when we get to four, clap and say, "Break."

As you and the children count together, break off lengths of four. As you break off the lengths of four, be sure you line them up one right under the other so that if you were to snap the lengths back together again, you would get the original pattern.

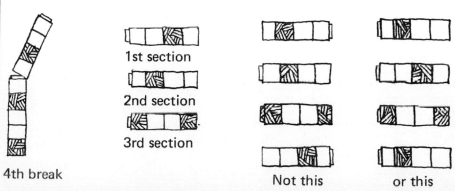

1st section

2nd section

3rd section

4th break

Not this   or this

Keep breaking off lengths of four until the pattern that forms is obvious (that is, when you can predict what will happen even before you place the next group of four cubes).

2   3   4   5   6   7   8

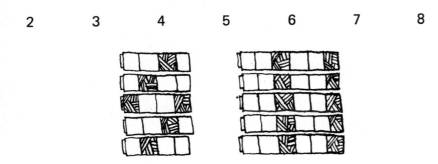

Have the children choose another number to break the train into. If a child says "Six," ask the children if they think the six pattern will look the same or different from the four pattern.

As the children count with you (clapping when it is time to break the train off), line up the lengths of six and look for a pattern. Tell the children to call out "I see a pattern" when there are enough cubes so that the pattern is obvious.

2   3   4   5   6   7   8

Compare the two patterns. Are they alike or different?

Choose another number. Predict whether it will look like the four or the six or different from either.

Let's try seven. I think it will look different from all the rest.

Continue to break the train into the other lengths. Discuss the patterns that emerge. Predict what would happen with other numbers.

2   3   4   5

6   7

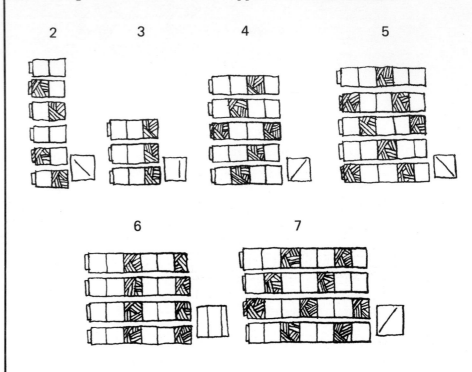

The relationship between the patterns will be easier to see if you draw the lines formed by one of the colors (in this case, green).

On succeeding days, explore other patterns with the children.

For example:

2   3   4   5

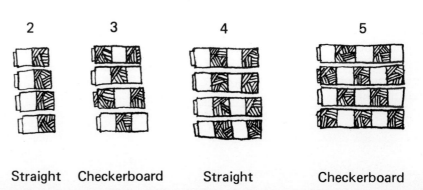

Straight   Checkerboard   Straight   Checkerboard

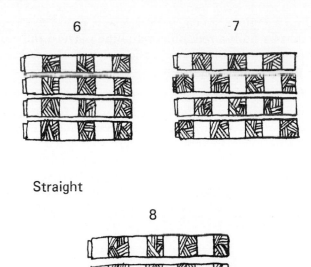

6   7

Straight

8

00	01	02	03	▣	05	06	07	▣	09
10	11	▣	13	14	15	▣	17	18	19
▣	21	22	23	24	25	26	27	28	29
30	31	32	33	34	35	36	37	38	39
40	41	42	43	44	45	46	47	48	49
50	51	52	53	54	55	56	57	58	59
60	61	62	63	64	65	66	67	68	69
				74		76	77	78	79

On another day the child picks a different number, places a cube on that number, and then counts and places cubes on the chart to determine a different pattern. For example:

00	01	02	▣	04	05	▣	07	08	▣
10	11	▣	13	14	▣	16	17	18	19
20	21	22	23	24	25	26	27	28	29
30	31	32	33	34	35	36	37	38	39
40	41	42	43	44	45	46	47	48	49
50	51	52	53	54	55	56	57	58	59
60	61	62	63	64	65	66	67	68	69
				74		76	77	78	79

Extension: The children can look for patterns and see if any of the numbers go with the pattern strips that are hanging in the room. In the above example, all the straight patterns fit the 2, 4, 6, 8 pattern.

PATTERNS ON THE 00–99 CHART

This activity gives children the opportunity to encounter some of the number patterns they may have experienced in other settings and to discover new patterns.

The child picks a number from two to ten and places a yellow cube on that numeral on the 00–99 chart (see black-line master 70). If the child picks four, for example, she or he places a cube on the numeral 4. Then, starting with the space next to the four, he or she counts to four over and over, placing a yellow cube on every fourth square.

For example:

Extension: If you have labeled patterns with colors and hung pattern strips in the room as described on page 46, have the children list the numerals they covered with the cubes and check the strips to see what color their pattern is. If there is no strip to match a child's pattern, a new color can be chosen and the pattern added to the others.

Variation: Have the children do the same activity using grids other than the 00–99 chart.

For example.

00	01	02	03	▣
05	06	07	▣	09
10	11	▣	13	14
15	▣	17	18	19
▣	21	22	23	24

00	01	02	03	▣	05	06	07
▣	09	10	11	▣	13	14	15
▣	17	18	19	▣	21	22	23
24	25	26	27	28	29	30	31
32	33	34	35	36	37	38	39
40	41	42	43	44	45	46	47

ANALYZING AND ASSESSING YOUR CHILDREN'S NEEDS

The concept of pattern is not something you can check off as having been mastered in the same way as the skill of recognizing numerals 1 through 5. Pattern can be understood at different levels. If you assign children the same task at the same time, a wide range of levels will be displayed.

The following children have all been assigned the task of making up their own patterns with Unifix cubes.

Joe is building a yellow, green, yellow, green pattern. His teacher remembers that he built a red, blue, red, blue pattern the day before. In fact, it appears that whenever Joe is assigned the task of making patterns, he builds only AB patterns.

The AB pattern is one that many children lock into when beginning to work with pattern. Joe seems to have identified the word *pattern* with AB patterns. His teacher needs to help him become more flexible and aware that pattern can be made in many different ways, as long as you can predict what comes next. To help him move away from AB patterns, the teacher can simply suggest such things as "Can you make a pattern with two greens touching?" or "Can you make a pattern with three colors?"

◆ ◆ ◆

Peter snapped four cubes together (blue, blue, red, green) and said "I made an AABC pattern. Now I'm going to make an AABBCC pattern."

Peter's teacher wants him to realize that patterns are not just various colors joined together and labeled. The important thing about pattern is that it can go on and on, and Peter's teacher wants him to experience that. She tells him, "I want you to make your pattern very long. Can you make it go all the way from here to the table? (or as long as your body? or across the whole table?) I'll come back in a few minutes and see how long you have made it."

◆ ◆ ◆

Leticia and Scott are randomly snapping the cubes together. They do not seem to have extracted the idea of pattern from the group experiences.

Their teacher says, "Watch me make a pattern." (He builds red, red, red, red.) "What do you think is next? and next? and next? and next? Let's try another one." (He builds green, green, green, green.) "What do you think is next? Make it longer."

"Let's try this pattern. (The teacher builds red, red, green, green; red, red, green, green.) "What do you think is next? and next? Okay, now you make a pattern like this one."

Leticia immediately begins to make a red, red, green, green pattern.

The teacher will keep a special eye on Leticia in the days to come. He may need to help her get started for a few days. He will want to make sure that Leticia doesn't just repeat the pattern she did successfully with the teacher today. He will give her various task cards to copy and extend and, after some experiences with these, will again encourage her to use her own ideas.

Scott uses red and green cubes but just uses whatever color is handy and does not make a pattern. Scott does not yet see the underlying organization of pattern. The teacher will continue to provide opportunities for Scott to experience patterns. He will show him various patterns and have him name the colors (such as yellow, yellow, red; yellow, yellow, red; yellow, yellow, red; etc.), and perhaps over time Scott will begin to notice the repetition that occurs. The teacher will have him make trains that are all the same color so Scott can discover that he knows what comes next. He will build different kinds of patterns with him (such as up, down, up, down, up, down) to see if arrangement is more evident to him than color. He will give him task cards that he can copy directly, even though he may not yet be able to extend them. In other words, the teacher will continue to provide Scott opportunities to experience pattern in a variety of ways so that over time he will sort those experiences out in his own way.

◆ ◆ ◆

Carlene has made a long pattern. When she is describing it to her teacher, she finds she has a mistake in it.

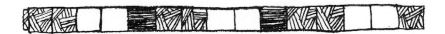

Red, red, white, white, black; red, red, white, white, black; red, red, white, white, . . . oops, I left out a black!

Carlene's teacher praises her for finding her own mistake. "Aren't patterns neat! We can tell what's wrong all by ourselves!" One of the things that makes patterning so powerful a tool is that mistakes become obvious. Children should be encouraged to feel good when they can find their mistakes.

◆ ◆ ◆

Suzanne has been making many complex patterns, and the teacher feels she is ready for a challenge.

Suzanne's teacher believes she is ready to see some of the number relationships that are in patterns. She poses questions such as the following.

Make a red, green, red, green pattern as long as the table. Count the reds. Guess how many greens are in your pattern, and then check to see how close you are.

Make an AAB pattern as wide as the table. Count the color cubes you used for the A part of your pattern. Estimate how many B cubes you used. Check and see how close you came.

Think of other patterns. Count one of the colors. Guess how many of the other colors are in the pattern you made.

◆ ◆ ◆

Children can work successfully with pattern at many different levels. In most cases it will not be necessary to keep detailed records of children's progress in pattern; however, the following outline may help you analyze the level at which a child is working if you feel you need specific information.

Can the child:

Copy patterns?
Extend patterns?
Describe patterns with words? with ABC's?
Build a pattern when given a rhythmic pattern? ABC's?
Create his or her own patterns?
See patterns in series of numbers and make predictions? using a 00–99 chart? without the help of a chart?

When answering the above questions, make note of the type of pattern the child is or is not having success with. Is she or he working with simple or complex patterns? Generally, the AB, ABC, AAB, or AABB patterns are simpler for children to handle. The more complex patterns are ones such as ABACAD, ABBCD, or AABAACAABAAC.

CHAPTER THREE

If your textbook or workbook objectives
are: • Finding the larger and smaller numbers
        • One more than, one less than
        • Greater than >, less than <

Then you are dealing with:

The Concepts of More and Less

Section I.   Developing and Extending the
              Concepts of More and Less
Section II.   Connecting Symbols to the
              Concepts of More and Less

WHAT YOU NEED TO KNOW ABOUT THE CONCEPTS OF MORE OR LESS

To adults, the concepts of more and less seem like they should be obvious or simple for children to grasp. They're not. Consider the concept of conservation of number discussed in Chapter One. Before children have developed conservation, they believe that a group of objects when spread out is more than when that same group of objects is pushed together. For example, if the teacher shows a child some cubes lined up in the following manner and asks "Are there more red

(red)

(blue)

ones or more blue ones?" children will usually answer easily that there are more blue ones. However, if the teacher spreads the red ones apart

and asks, "Now are there more red ones or blue ones?" young children will often answer that there are more red ones. Even counting does not convince children at a certain stage of thinking that the two groups are the same.

Another example of children's perceptions of things is apparent in the following situation.

A group of children were finding out how many of their footsteps it took to go from one edge of the rug to the other.

Three, four . . .

When the teacher asked if they thought it would take more or less of her footsteps, the children shouted out, "More, 'cause your feet are bigger."

The concepts of more and less can be confusing and elusive, because they are relative ideas: sometimes five is more and sometimes five is less. Five is more when compared to three, but less when compared to ten.

In order to sort this all out, children need lots of experiences comparing quantities. They need to begin simply, determining which of two quantities is more and which is less. When this idea is established, then they need experiences that will help them develop a stronger sense of the relationships between numbers by focusing on how many more or less one number is than another.

It is important to keep in mind as you offer these experiences to children that they learn these relationships at different rates of speed and at different times. The fact that a child doesn't know six is two more than four or five is two less than seven does not indicate that you need to drill him or her until this information is learned at a rote level. It is your responsibility, rather, to provide children the opportunities to explore these relationships in ways that allow them to count and find out what they don't already know. Through these kinds of experiences, over time, they will learn what you want them to learn.

When you work with young children, you will see that the concept of less is much more difficult for them to deal with than the concept of more. Thinking about what is not there is harder than thinking about what is there. The activities from this chapter provide opportunities for the children to try and figure out how many less one number is than another. Give them whatever help is necessary as they are learning.

Section I: Developing and Extending the Concepts of More or Less

These lessons should not be considered as a unit that you do for a certain period of time before you move on to other concepts. Integrate these activities with others in your math lessons throughout the year, raising the level of difficulty of the activities (as described when each is presented) to meet the needs of the children.

Teacher-Directed Activities

The following activities should be experienced over and over again. They can be presented at different levels, according to the needs of your children. Some children will need to begin by focusing on the general concept of more and less and will need to work with that concept (along with other number concepts) for many weeks. Other children will need to begin with the general notion of more and less but will be ready to move relatively quickly to exploring how many more or less one number is than another. Other children who already understand the concepts of more and less will be ready to begin the work with the activities in this chapter for determining how many more and less one number is than another.

You will be able to determine which children are ready for a particular level by observing the responses as you present the activities to them. (See "Analyzing and Assessing Your Children's Needs.")

MORE AND LESS SPIN IT

Materials: Unifix cubes • More/less spinner—see p. 217 for directions for making • Unifix Number Indicators (optional—see p. 217)

Tell the children to build a Unifix tower of a particular height.

For example: *Build a tower that is 5 cubes high.*

The children each take a turn spinning the more/less spinner to determine if the group will build towers that are more or less than the original tower. (Have them label their towers with Unifix Number Indicators, if available.)

For example: If the spinner landed on less, the children would build towers less than five. Each child holds up his or her tower and tells how high it is. The group then describes the relationship.

My tower is two cubes high. Group: Two is less than five.

My tower is four cubes high. Group: Four is less than five.

If the spinner landed on *more,* the children would build towers more than the original tower of five cubes.

My tower is nine cubes high. Group: Nine is more than five.

My tower is seven cubes high. Group: Seven is more than five.

(Repeat, starting with towers of different heights.)

Extension: Have the children tell how many more or less their tower is than the original one.

My tower is two cubes high. Two is three less than five.

STACKS

Materials: Unifix cubes • Unifix Numeral Indicators (optional)

The children each make a train of ten to twelve cubes and hide them behind their backs.

When the teacher says the signal *stacks,* the children each break off a part of their trains and place the parts in front of them. At the same time, the teacher also places a stack of cubes in front of him or her.

As the children each take a turn holding up their stacks, the group compares it to the teacher's stack. (The group responds rather than the individual child so no one child is put in a testing situation.)

For example: Two is less than three.
Five is more than three.
Three is the same as three.

Extension: The children tell how many more or less their stacks are than the teachers. It will help if the child holds the train right next to the teacher's train.

Five is two more than three.

IS IT MORE OR IS IT LESS?

Materials: Unifix cubes

Direct the children to build two trains of specific lengths. Then compare the trains to see which is more and which is less.

For example:

Build a red train that is seven cubes long.
Build a blue train that is three cubes long.
Show me the train that is less.
Show me the train that is more.

Variation: Roll a die to determine the length of the two trains.

Extension: Ask the children to tell how many more or less one number is than another.

How many more is seven than three?
How many less is three than seven?

GRAB BAG

Materials: Unifix cubes (2 colors) · Grab bag

A child takes a handful of cubes from a grab bag (or two handfuls if her or his hands are too small to take as many as you need them to take). The cubes of each color are snapped together, and the group compares them to see which color is more and which is less.

Jamie got two red cubes and six green cubes. Two is less than six. Six is more than two.

Extension: Ask the children to tell how many more and how many less one number is than another.

SAMPLES

Materials: Unifix cubes (2 colors) · Grab bag

Set up a grab bag that has twice as many cubes of one color as another (the children should not know what is in the bag). Have each child in the group take one cube out of the bag. Snap together each color, and determine which color was picked more often. Using this information, predict which color cubes there are more of in the bag. Dump the cubes out of the bag to check.

For example: Put in ten red and twenty blue cubes. Have each child pick one cube.

I got a red.
Mine is blue.
I got blue, too.
Another blue!
Look, I got a red.
Blue.
Blue, again.
Mine is red.

We got five blue and three red.
We think there are more blue than red.
Let's check and see.

Repeat to see if the results are the same.

Variations: Set up a variety of situations:
a. Equal numbers of each color
b. Just a few cubes of a second color
c. Use three colors

SPIN AND PEEK

Materials: Unifix cubes · 9 margarine tubs · More/less spinner—see p. 217 for directions for making

This game is similar to the traditional game of "Concentration." In this version the children search for groups that are more or less than other groups. Hide various quantities of cubes under the tubs. The children take turns lifting a tub to see how many cubes are under it. They then turn the spinner to see if they should look for a group that is more or less than the group under that tub. If they are successful, they remove both tubs and the cubes under them. If not, they replace them.

For example:

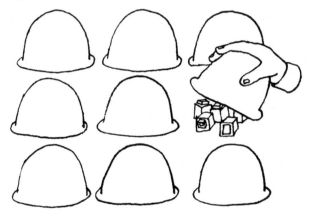

There are five cubes under this tub.

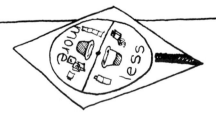

The spinner landed on less. I need to find a tub with less than five under it.

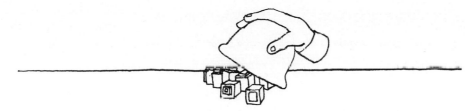

No. This tub has seven under it.

The next child takes a turn.

There are three cubes under this tub.

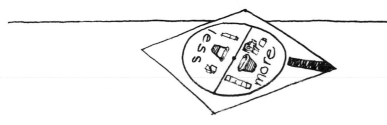

The spinner landed on more. I need to find a tub with more than three cubes.

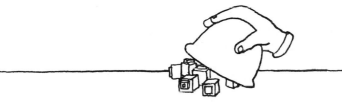

This tub has six. That's more than three. I can take these two tubs out of the game.

Continue until all the possible matches have been made.

COUNTING BOARDS

Materials: Unifix cubes • 1 counting board per child—see p. 210 for directions for making

Tell stories, and have the children act them out.

There are four cows in the field. There are more horses than cows.

Have the children tell what they did to act out the story.

I put seven horses in the field, because seven is more than four.
I put nine horses in the field. Nine is more, too.

Extension: Tell them stories that require them to determine numbers as in the following example.

There are two cows in the field.

There are two more horses than cows.

Show me the cows and the horses.

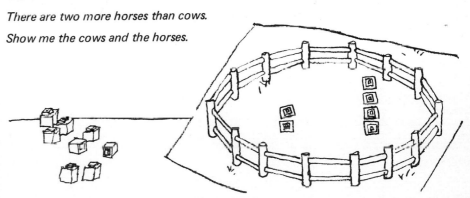

(This task can be kept easier if you use numbers that are just one more or one less than the other.)

UNIFIX CUBE GRAPHS

Materials: Unifix cubes

Provide children opportunities to compare numbers less than ten by having them make graphs with Unifix cubes as part of their small-group instruction time. (You can extend this activity to larger numbers if you involve the whole class in making the graphs.) A graph is a visual representation of information that highlights relationships. The Unifix cubes serve this purpose very well. To make a Unifix cube graph, (1) ask the children a question, (2) have them choose the specified cube to represent their answers, (3) snap together the cubes, and (4) discuss the results.

For example:
Ask a question.

Do you have a sister?

Direct students to get a cube.

If you have a sister, get a red cube. If you do not have a sister, get a blue cube.

Organize the information by snapping the cubes together.

Snap together the red cubes and the blue cubes to compare.

Discuss the results. Ask the following questions.

How many people in our group have sisters?
How many people in our group do not have sisters?
Are there more people with sisters or more people without sisters?
Are there less people who have sisters or less people who do not have sisters?
How many more have sisters than do not have sisters?
How many less do not have sisters than have sisters?

After many experiences with a variety of graphs, having asked the previous types of questions, ask the children to tell you what they can see when looking at a graph.

For example: Did you choose a square cracker or a round cracker at snack time today? Get a yellow cube if you chose a square. Get a green cube if you chose a round cracker.

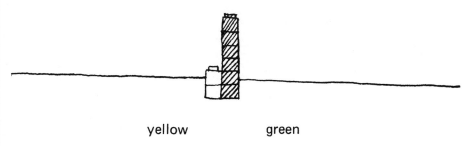

yellow   green

What can you tell when you look at our graph?

More people chose round crackers.
Less people picked squares.
Six people picked round crackers.
Two people picked square crackers.
There were four more people who picked round crackers than square crackers.

Suggested graphing questions to ask the children:

- Are you wearing buttons?
- Did you walk or ride to school this morning?
- Does your name have an "a" in it?
- Do you like french fries or potato chips better?

- Have you ever had gum in your hair?
- Does your name start with a consonant or a vowel?
- Did you bring your lunch today?
- Do you have a dog?
- Are you right-handed or left-handed?
- Which do you like best—applesauce or apple pie?

BUILD A STACK

Materials: Unifix cubes

Give the children a variety of directions, requiring them to use the concepts of more and less to figure out the stacks you want them to build.

For example:

Build a stack that is one more than four.
Build a stack that is two more than six.
Build a stack that is one less than seven.
Build a stack that is two less than three.

Children will respond to these directions in many ways. If they appear completely bewildered, say

Build a stack of four.
Now build a stack that is one more than four. What number did you make?

Build a stack of six.
Now build a stack that is one less than six. What number is it?

Watch to see how they solve the problems. Do they build second stacks or do they add or take away from the original? If they are confused, model what you want them to do.

If the children are doing the task easily, see if they can predict the length of their stacks before they build them.

For example:

What number do you think is one more than five? Let's build it and see.

NUMBER CARDS

Materials: Unifix cubes · Number cards (teacher-made; see examples below)

Show the children two cards with numerals written on them.

Tell them to build the larger number.

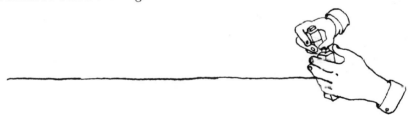

Ask them what they have to do to make the smaller number.

We have to take some off.

Yes, we have to take four off.

Encourage them to take the cubes off as a unit (i.e., still joined together).

Show two different cards.

Tell the children to build the smaller number.

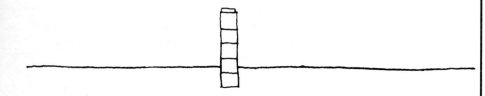

Ask them what they have to do to make the larger number.

> We have to
> put two more on.

Variation: Build both trains. Have the children tell which is more and which is less.

Extension: Ask the children how many more or less one train is than another.

GROW AND SHRINK

Materials: Unifix cubes • Working space paper—see p. 210

Call out a number, and the children will place that number of cubes on the working space paper.

For example:

Six.

Call out another number. Ask the children if the number is more or less than the number on their working space paper.

Eight. Is that number more or less than six?

> More.

Have them make the number and tell *how many* more or less it is.

How many more do you need to get?

> Two more.

Continue calling out other numbers and having the children determine if they are more or less and then how many more or less.

Five. Is that more or less than eight?

Less.

Five is how many less than eight?

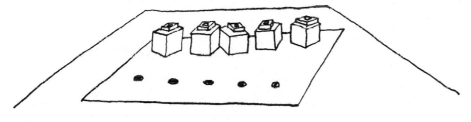

Three.

THE MAGIC BOX*

Materials: Magic box and Magic box cards—see p. 213 for directions for making • Unifix number lines (1 per child)—use number train outlines on black-line master 32; cut apart and mount on tagboard; number from 0 to 14; laminate or cover with contact paper • Unifix cubes

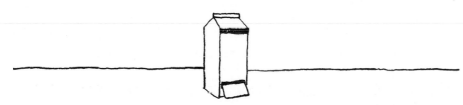

The magic box cards are sets of cards that have a particular number relationship written on the front and back. (That is, the number on the front of a set of cards is a certain number more or less than the number on the back.)

For example, one set of cards might have the following numbers:

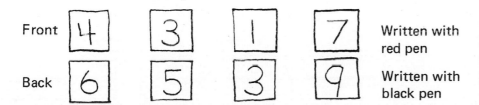

Front	4	3	1	7	Written with red pen
Back	6	5	3	9	Written with black pen

As you can see, each number on the back is two more than the number on the front. Another set of cards might have the following numbers:

Front (red)	7	4	5	2	etc.
Back (black)	6	3	4	1	

What number relationship do these cards have?

The magic box is designed so that when a card is placed in the top slot, it will turn over so that the number on the back appears when the card comes out the bottom slot—like magic!!

Have a child choose a magic box card from the set you have provided and show it to the rest of the group. Each child places that number of cubes on their number lines.

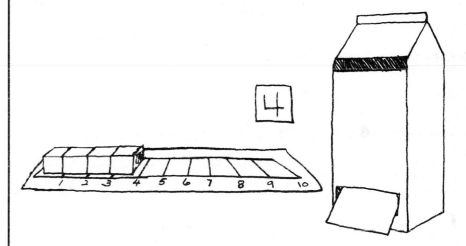

The child puts the magic box card into the top slot of the magic box and then shows the group the number on the card when it comes out the bottom slot. Each child then adds or subtracts cubes until they reach the number indicated on the card.

*Based on MATHEMATICS *THEIR* WAY, p. 248.

We put two more cubes on to make six. Six is two more than four.

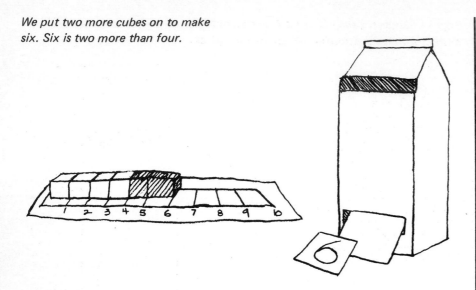

Another child chooses a magic box card and shows the number to the rest of the group. Each child builds that number on their number lines.

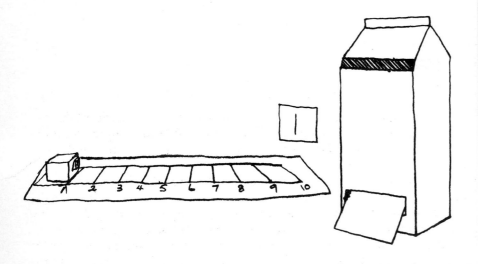

The child then puts the card into the magic box and shows the rest of the group the number that comes out. The children again add or subtract cubes until they reach the number indicated by the card.

We put two more on to make three.
Three is two more than one.

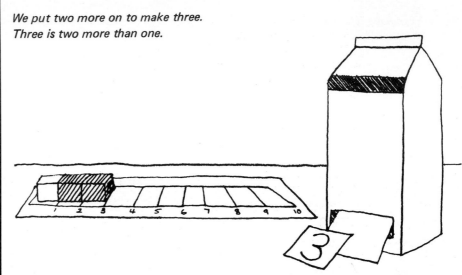

After three examples, ask the children to predict what number they think will come out of the magic box. In the examples shown above, the number on the back of the card is two more than the number on the front of the card, which many children will be able to predict. Be sure you actually check each prediction. On succeeding days, repeat the activity using a different set of cards so the children will have new patterns to discover.

For example:

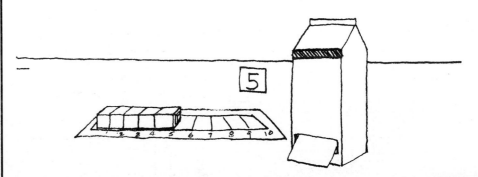

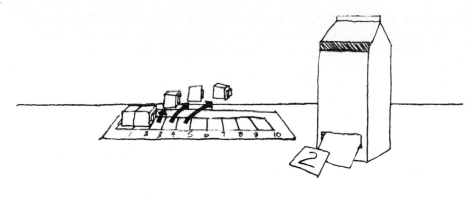

We took three off to make two.

Continue with cards that repeat the pattern of taking three off.

Variation: Using black cubes, have the children build the Unifix train indicated by the black number on the card.

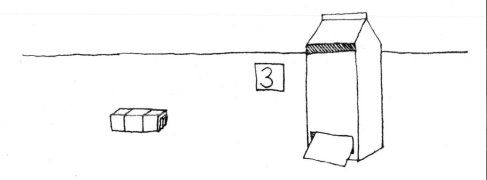

Place the card in the magic box, then build the red number that comes out of the box, using red cubes. Compare the red and black trains.

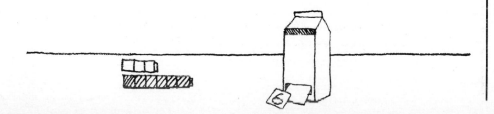

Repeat using different cards. After the children see the number relationship, have them predict the red number.

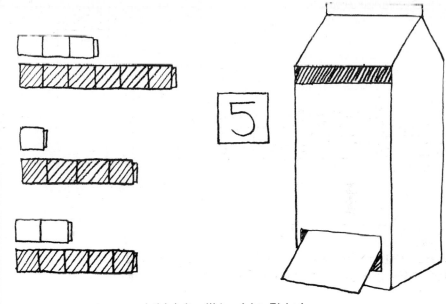

I think it will be eight. Eight is three more than five.

Independent Activities

Children can do these activities alone or with partners, without direct teacher involvement. They have been arranged from simple to more complicated so you can meet the changing needs of your children throughout the year.

Many of the activities will need to be modeled in a teacher-directed setting before you put them out as independent work. Teach the children how to play a game by going through it step by step, using one child as your partner. After the initial introduction, have the children tell you and a partner what to do for each step of the activity. Model the game for the group for the next two to four days, until most of the children know what to do and can help those children who don't.

HANDFULS—A Game for Partners*

Materials: Unifix cubes (2 colors) • 2 pieces of paper (9 x 12) • More/less spinner—see p. 217 for directions for making • 2 scorecards (3 x 5 cards with *Score* written on them)

Partner A takes two handfuls from a pile of Unifix cubes of two colors mixed together. She or he places them on a piece of paper and covers them with another piece of paper.

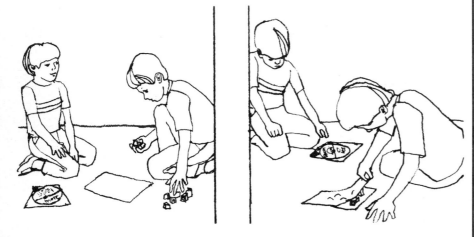

Partner B turns the spinner and notes if it lands on more or less. Partner A lifts the paper briefly and allows Partner B to get a quick look at the cubes. If the spinner landed on more, Partner B guesses which color cubes there are more of. If the spinner landed on less, he or she guesses which color is less.

Partner B: I think there are less blue.

*Based on MATHEMATICS *THEIR* WAY, p. 125.

Partner B keeps track of the guess by taking a Unifix cube from the pile (not off the paper) and placing it on the appropriate side of the spinner. For example, if she or he guesses less blue, he or she would place a blue cube on the less side of the spinner.

Partner B then snaps together the cubes of one color (blue in this example, because that's the color she or he picked), and Partner A snaps together the cubes of the other color. They compare them to see which color is more and which is less. If Partner B guessed right, he or she takes the cube from the spinner and places it on his or her scorecard. If she or he guessed wrong, she or he puts it back in the pile.

There are three blue cubes. That's less than six orange cubes. I can put one cube on my scorecard.

The partners then switch roles; Partner B now takes the handfuls and hides them, and Partner A turns the spinner and guesses. When the time is up for playing the game, the partners snap together the cubes on their scorecards and compare to see who has more and who has less. The person with more does not necessarily have to be the winner. The partners can turn the spinner one more time. If it comes up more, the person with more cubes wins the game. If it comes up less, the person with less wins the game.

It landed on more. I win the game.

STACK, TELL, SPIN, AND WIN——A Game for Partners*

Materials: Unifix cubes • More/less spinner—see p. 217 for directions for making

Each partner begins the game with a Unifix train the same length as the other (about twenty cubes long).

Accomplishing this task is an interesting first step that can be handled in a variety of ways. One way is to have about forty cubes in a baggie available to the children when they play this game. They can snap the cubes together, comparing lengths until they end up the same. That can be a valuable problem-solving situation in itself. Another way to handle the task is to provide a string that is the appropriate length, and have the children each build a train as long as the string. A third way is to have a line drawn on the chalkboard that indicates the appropriate length.

Once they have identical trains, the children sit facing each other, putting their trains behind their backs.

With their hands behind their backs, the partners simultaneously break off a piece of their trains (stacks) and place those pieces in front of them.

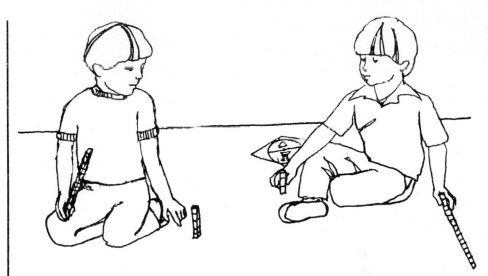

Each child tells the other if his or her stack is more or less than the other person's stack. (If the stacks are the same, they put those back and put out different lengths.)

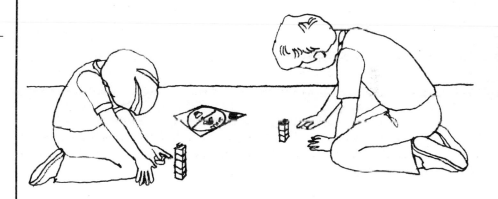

Mine is more.   Mine is less.

One partner turns the spinner. If it comes up on more, the child whose stack was longer takes both stacks. If the spinner lands on less, the child with less takes both stacks.

*Based on MATHEMATICS *THEIR* WAY, p. 126.

It landed on less. I get to keep
both stacks.

They continue to play until one player runs out of cubes.

COMPARING UNIFIX PUZZLES

Materials: Unifix puzzle cards—see p. 211 • Unifix cubes • More/less and same
cards—see p. 212 for directions for making

The child picks two Unifix puzzle cards. She or he predicts which
puzzle will take more cubes. After placing the appropriate number of
cubes in each puzzle and determining how many cubes fit in each, the
child labels the puzzles with the more/less cards.

I think this will be more.

The child continues comparing other puzzles in the same way.

COMPARING CONTAINERS

Materials: Unifix cubes • A variety of small containers (boxes, jars, margarine tubs,
etc.) • More/less and same cards—see p. 212

The child chooses two small containers and determines which of the
two holds more and which holds less than the other. He or she labels
each container with the appropriate more/less card. Repeat, comparing
different containers.

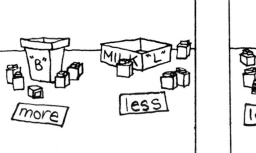

This one has more, L has less.

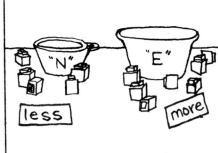

E has seven cubes. N has five cubes.
E is more; N is less.

COMPARING LINE PUZZLES

Materials: Unifix cubes • Line puzzle task cards—see p. 212 • More/less and same cards—see p. 212

The child chooses two line puzzles. After placing the appropriate number of cubes on each one, she or he labels them with the appropriate more/less card. The child repeats the activity, using a variety of puzzles.

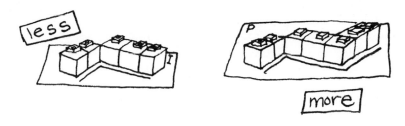

I counted six on the I card.   I counted eight on the P card.

SORT AND COMPARE

Materials: Unifix cubes (2 colors) • Various small containers (boxes, jars, margarine tubs, etc.) • More/less and same cards

Fill small containers with two colors of Unifix cubes. The child takes a small container and determines which color is more and which is less. He or she labels the cubes with the appropriate more/less card.

Extension: Put three colors in the containers, and have the children snap together the cubes of the same color and put them in order.

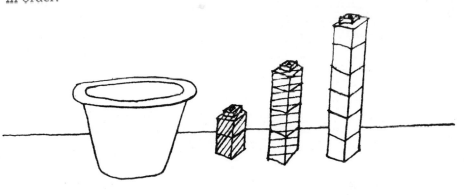

COMPARING NUMBERS—A Game for Partners

Materials: Unifix cubes • Small number cards • More/less spinner—see p. 217

Each player draws a numeral card from a pile of cards.

Partner A   Partner B

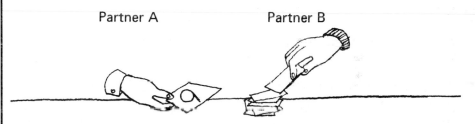

They each build the appropriate Unifix stack to match their cards.

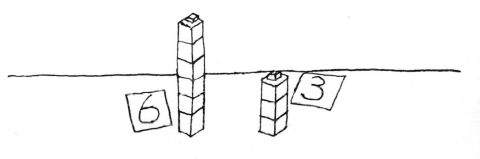

One child turns the more/less spinner. If it lands on less, the partner whose stack has less cubes wins both stacks. If the spinner lands on more, the partner with more cubes wins both stacks.

Partner A   It landed on more. I win.

They continue to take turns, each child accumulating Unifix stacks.

When the time is up or the children decide to end the game, they snap together all the stacks of cubes. They turn the spinner to see if the person who has accumulated more or less cubes is the winner.

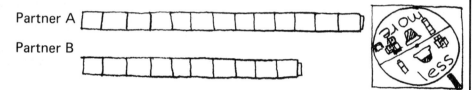

Partner A

Partner B

Partner A   I have less. I win the game.

COMPARING COMBINATIONS—A Game for Partners

Materials: Unifix cubes (2 colors) • Addition equation cards—see p. 214 • More/less spinner

Each partner draws an equation card from a pile of cards.

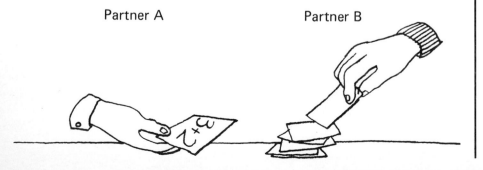

Partner A   Partner B

Using two colors, each player builds the appropriate Unifix stack to match their cards.

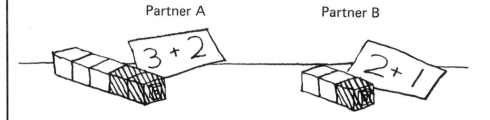

Partner A   Partner B

One child turns the more/less spinner. If it lands on less, the partner with less cubes in the Unifix stack wins both stacks. If the spinner lands on more, the partner with more cubes wins both stacks.

Partner B   It landed on less. I won!

They continue to play in the same way, each child accumulating stacks according to the spinner.

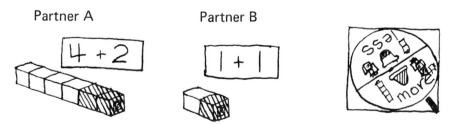

Partner A   Partner B

Partner A   It landed on more. I won!

When the time is up or the children decide to end the game, they snap together the stacks they have won and compare them. They turn the spinner one last time to see if the player who has accumulated more or less cubes is the overall winner.

Partner A

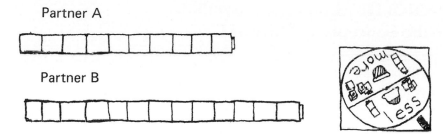

Partner B

Partner A   It landed on less. I won!

Extension a.: Use both addition and subtraction cards.

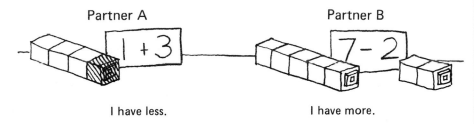

Partner A

I have less.

Partner B

I have more.

b. Use addition and multiplication cards.

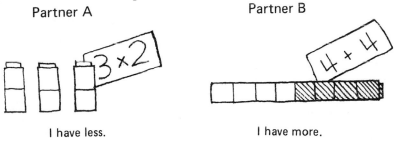

Partner A

I have less.

Partner B

I have more.

APARTMENT BUILDINGS—A Game
for Partners or Groups of Three or Four

Materials: Game board for each child—black-line master 33 · Dice (0–4) · Unifix cubes

Each child builds apartment buildings (Unifix stacks) on each "lot" on her or his game board. Each completed apartment building is to have

the same number of stories, and each building under construction must be completed before a new building can be started. The children take turns rolling dice to determine the number of stories to be added to their buildings at each step. The child who completes all his or her apartment buildings first is the winner.

For example: The children have decided that each building is to be four stories high for this game. (A different number could be picked when playing the game at a different time.)

Partner A

I rolled a three. My first building has three stories so far.

Partner B

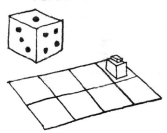

I rolled a one. I can build one story of my first building.

Partner A

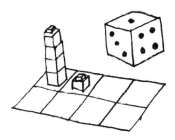

I rolled a two. I need one of the two cubes for my first building. I can start a new building with the other cube.

Partner B

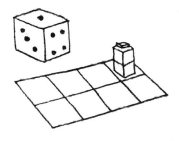

I rolled a one again. I can put another story on my building. I need two more cubes to finish my building.

Partner A

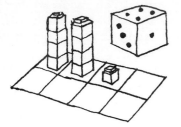

I rolled a four. I can put three stories on my second building, and I have one more to start my next building.

Partner B

I rolled a three. I can put two on my first building. I have one more to start my next one.

The partners continue taking turns and building apartments until one partner has completed all the apartments she or he can and wins the game.

Note: This game is very effective but somewhat complicated. Most children will need the game modeled several times before being able to play it independently. It can be played as a small-group, teacher-directed activity, with each child in the group constructing buildings on his or her own gameboard. Rather than racing to complete the buildings, the children simply practice building them according to the rolls of the dice.

For example: Each apartment is to be six cubes high.

Lisa rolled a three. Everyone make three stories on your apartment buildings.

Ray rolled a four. Everyone get four cubes. How many of the cubes do we need to finish our first building? How many will we have left to start our next?

Section II: Connecting Symbols to the Concepts of More or Less

The purpose for the activities in this chapter is to help children develop a stronger sense of number and number relationships, which will help them in all their encounters with number ideas. The concept is more important than the use of the greater than ($>$) and less than ($<$) signs.

If you must teach your children the greater than and less than signs, the following activities will provide the opportunity to bring some meaning to these symbols and are appropriate only after the children understand the concept. It is recommended that you *not* teach these symbols until after the children are at ease with addition and subtraction equations (see Chapter Four).

Teacher-Directed Activities

You can introduce the children to the symbols for greater than and less than by recording the various number relationships that occur when doing the activities in Section I. The following steps can be helpful no matter which of the activities you are working with.

Step 1. The teacher writes the appropriate symbols as the children tell what numbers they are comparing.

For example: If you are playing "Stacks" with your children and you have put out a stack of five and a child has put out a stack of two and says, "Two is less than five," you write $2 < 5$.

If another child has a stack of seven and says, "Seven is more than five," write $7 > 5$.

If the occasion arises and two stacks of the same number are compared, write the appropriate equation $5 = 5$.

Step 2. The children copy the symbols on individual chalkboards as the teacher writes.

Step 3. The children write the symbols before the teacher does when two numbers are compared. After the children have written, the teacher writes the symbols so the children can check.

There are a variety of ways that teachers have used to help their children remember which way the symbols go. The following are some of the ways.

Imagine the > is an alligator's mouth. The alligator is always hungry, so he always eats the bigger number.

The > sign has a large opening at one end and a tiny point at the other end. The tiny point always points to the smaller number.

The signs are up in the classroom with appropriate clues so the children can copy them until they learn them.

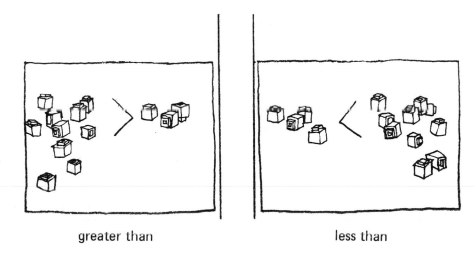

greater than   less than

Independent Activities

The following activities provide opportunities for the children to work with various numbers and record the relationships between them. They should be able to write the symbols >, <, = easily before being asked to work independently. Be sure the signs are posted so the children can refer to them if needed.

ROLL AND SPIN

Materials: Unifix cubes • More/less spinner (see p. 217) • Dice • Worksheets—see black-line master 31

Each child rolls a die and builds the appropriate trains.

For example:

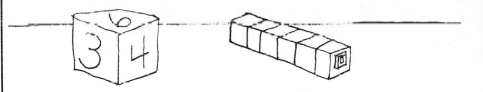

I rolled a six.

They turn the spinner. If it lands on less, they build a train that is less than the original number and record the appropriate number sentence.

If the spinner lands on more, they build a train that is more than the original number and record the appropriate number sentence.

Repeat several times.

I rolled a four this time.

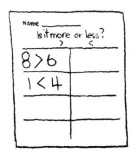

The spinner landed on less. I need
to build a train that has less than four.

MORE OR LESS

Materials: Unifix cubes • Dice • Worksheets—see black-line master 31

The children are assigned or are allowed to choose a particular number,
and they build a train that length.

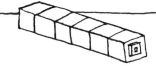

I'm going to work with six.

A die is rolled to determine a second train to build.

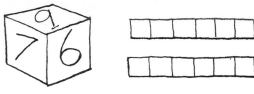

I rolled a nine.

The child compares the number rolled to the original train and writes
the appropriate number sentence.

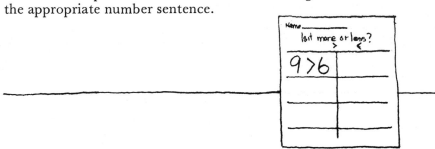

The die is rolled again, and another train is built and compared to the
original train of six. The appropriate number sentence is written.

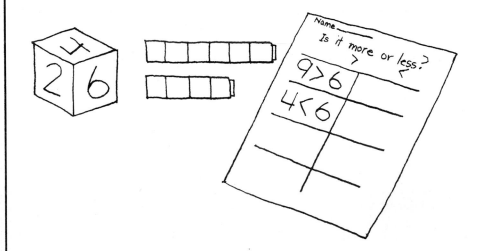

I rolled a four.   etc.

GRAB BAG

Materials: Unifix cubes (2 colors) • Grab bag • Worksheets—see black-line master 31

The child grabs a handful (or two) of cubes from a bag containing two colors of cubes.

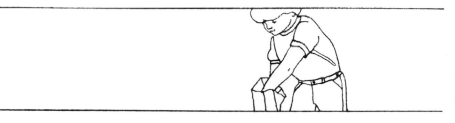

She or he snaps together the cubes that are the same color and compares them.

I got three blue and six yellow.

The child then writes the number sentence that describes the results.

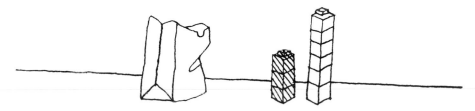

The child could write either 3 < 6 or 6 > 3. Both are correct.

COMPARING UNIFIX PUZZLES

Materials: Unifix cubes • Unifix puzzle cards—see p. 211 • Worksheets—see black-line master 31

The child picks two Unifix puzzle cards and determines the number of cubes it takes to fill each puzzle. He or she then writes a number sentence that describes the relationship. The child continues the activity, comparing other number puzzles in the same way.

COMPARING CONTAINERS

Materials: Unifix cubes • A variety of small containers • Worksheets—black-line master 31

The child chooses two small containers and determines the number of cubes it takes to fill each of them. She or he then writes the number sentence that describes the relationship. The child continues the activity, comparing other containers in the same way.

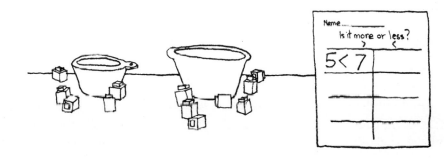

SORT AND COMPARE

Materials: Unifix cubes (2 colors) • Various small containers • Worksheets—see black-line master 31

The child takes a small container that has been filled with two colors of Unifix cubes and determines how many of each color is in the container. She or he then writes a number sentence to describe the relationship.

ANALYZING AND ASSESSING YOUR CHILDREN'S NEEDS

The understanding of number relationships develops gradually. If you periodically include activities in which the focus is on comparing numbers, children will develop a strong sense of number in their own time and way. The following descriptions of children and their responses to more and less activities will help you know what kinds of thinking are part of the normal development of these concepts. It is not generally necessary to do any in-depth, formal assessment. Just observe the children as you work with them.

LEARNING ABOUT LESS

Timmy had developed a good sense of the concept of more. He could tell easily what group of objects was more than another group of objects. The game "Handfuls" (p. 64) required Timmy to focus on and describe the group of objects that were less than another group of objects. As he groped for words to answer the question "Which of the two colors are less—the red or the blue cubes?" the best he could

come up with was "The blue ones have more less." In a roundabout way, that was the right answer expressed in a not-too-precise manner!

Donna was playing "More and Less Spin It" (p. 54) for the first time. She didn't know how to respond at all to the direction "Build a train that is less than five." She brightened up and busily made a train when the teacher said, "Build a train that is littler than five." When the teacher used the term *littler*, Donna could understand what was being asked of her. The concept was something she could deal with, but the term *less* had as yet no meaning for her.

Marlene had a worried look on her face. The teacher had told her to build a train that was less than five. She looked around at what the other children had done. Some of them had built trains of four, another had a train of two, another had a train of one. She just couldn't figure out what the "right" answer was. The teacher then took the trains that had been built and matched them to the five to show how they all were shorter than five, so all the trains were less and all the children were right. Marlene breathed a sigh of relief and said, "Now I get it. Lots of things are less."

All of the above children are just beginning to get the idea of both the concept of less and the language used to describe it. The teacher does not worry if they don't fully understand in one or two lessons. She knows they will deal with this same notion over and over again. There will be periods of time when children who seemed to understand one day will be confused the next. When comparing numbers, the "right" answers change all the time. What was less in one situation is more in another. Eight was less when compared to ten but was more when compared to six. Children may understand for a time, but they need to be reminded and to rethink what something as elusive as more and less means.

THE CONCEPT OF "HOW MANY MORE?"

A group of children working with their teacher have made a Unifix graph. The children who played in the sandbox at recess had chosen a red cube and those who did not had chosen a green cube. There were five red cubes snapped together and three green cubes snapped together. When the teacher asked, "How many more red cubes are there than green cubes?" the children answered that there were five more red cubes.

This is the common response for children who are just beginning to deal with the concept of "how many more." They are interpreting the question to mean "How many cubes in the stack with more?" The teacher needs to model for them what that language pattern refers to. She shows them with her hand where the two stacks are the same and what cubes tell how many more. With enough experiences over time, the children will be able to answer correctly on their own.

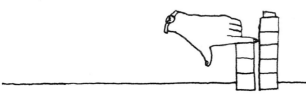

◆ ◆ ◆

Jamie can answer the question "How many more?" but needs to count the cubes to determine the answer. For example, there were eight red cubes and four blue cubes in two stacks lying side by side.

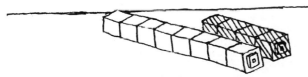

When the teacher asked, "How many more red cubes?" Jamie counted the cubes and said, "Four." On the other hand, Agnes has reached the point where she can tell without counting how many more for any number less than ten. When the teacher holds up two number cards such as 2 and 6, she answers immediately, "Six is four more than two."

Jamie is on his way to being able to do what Agnes can do. He simply needs more experiences. His teacher needs to make sure he has the experiences and is given the time necessary to reach Agnes's level.

CHAPTER FOUR

If your textbook or workbook objectives
are:
- Joining or separating sets
- Horizontal and vertical addition
- Horizontal and vertical subtraction
- Basic addition and subtraction facts

Then you are dealing with:

Beginning Addition and Subtraction

Section I.   Developing the Concepts of Addition and Subtraction

Section II. Connecting Symbols to the Concepts of Addition and Subtraction

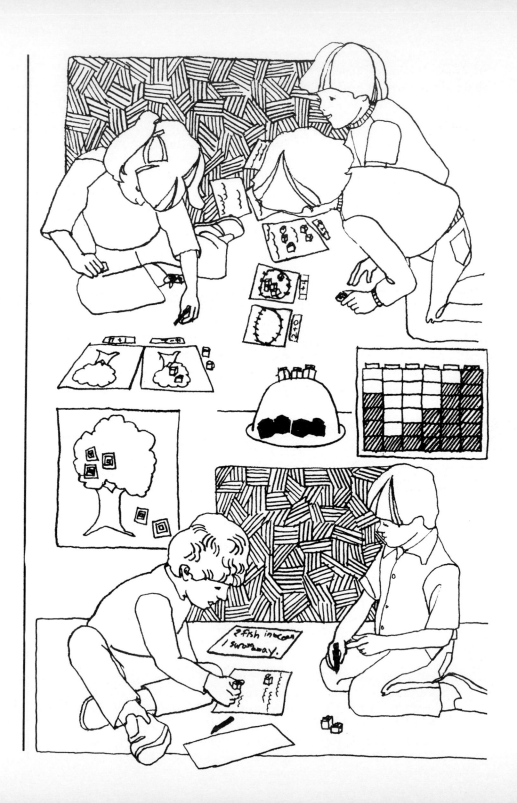

WHAT YOU NEED TO KNOW ABOUT BEGINNING ADDITION AND SUBTRACTION

The goal of most beginning arithmetic programs is that children learn the basic addition and subtraction facts. Success is often measured by how quickly the children can write answers to a series of problems. Teachers sometimes introduce children to addition and subtraction using some kind of manipulative materials; however, most teachers encourage the children to stop using these materials as soon as possible. The games that teachers use in an attempt to make the arithmetic period more enjoyable are designed to help children memorize the basic facts so do little more than provide additional work with symbols. In classrooms throughout the country, children spend most of their mathematics time practicing the basic facts by filling out workbook pages, using flash cards, and taking timed drill tests.

The results of all this time and energy expended by children and teachers are rather discouraging. It seems every teacher of young children has heard complaints from teachers of older children. Many children do not know the basic facts when they reach upper elementary grades. Even more children do not seem to know how to use their knowledge of basic facts appropriately when asked to solve problems. Arithmetic is a frustrating experience for teachers as well as children.

Some would say the problem is that children have not had enough drill. The basic assumption underlying this conclusion is that children can learn what is represented by a symbol by working with the symbol itself. But this assumption does not make sense. No one believes that learning to read, write, and spell the word *chocolate* is synonomous with experiencing chocolate; but many do believe that knowing how to read and write 3 + 4 = 7 is synonomous with understanding the number concepts represented by these symbols.

The following ideas must be understood by children if they are to deal with facts such as 3 + 4 = 7 in a meaningful way.

Addition and Subtraction—Real-World Processes. Children who deal with only symbols are misled into thinking that addition and subtraction exist as black marks on paper. Children need to know that adding is putting groups of objects together to determine how many and that subtracting is taking objects away to determine how many are left (or comparing groups to determine the difference between them). They need to learn that certain words and language patterns such as *total*, *sum*, *difference*, and *equals* are used in describing these processes that are familiar occurrences in their everyday lives.

Developing a Sense of Quantity. The subject of arithmetic is the description of quantities. The symbols represent *amounts* of things, but there is nothing inherent in a symbol that communicates how many that numeral represents. You may be able to get an idea of what that means if you put yourself in the place of a child for a moment. Instead of using the numerals you are already familiar with, we are going to use the letters of the alphabet in an unfamiliar way to help you understand what a child experiences when he or she is asked to deal with symbols before a sense of the quantities involved has been developed. Do not translate these letters into numbers and begin to think 1, 2, 3, because you will miss the insight into what children experience.

a = ●   b = ●●   c = ●●●   d = ●●●●   e = ●●●●● etc.

What letters belong in the following boxes?

c + e = □   b + d = □   d + f = □

How did you figure out the answers? Were your fingers useful? Now that you know the answers, do you think you could memorize the equations?

Given sufficient time and the threat of a report card or parent conference, you could memorize these facts and more. Would having memorized the above facts be of any help in answering the following questions? (Don't stop and figure out the answers. Just notice how knowing the equations above relate to the following questions.)

How much more is g than c?
Which is more? c + f or d + e?
If you had g people at a party, would s cookies be enough, too many, or just about right?
About how many grapes could you hold in one hand? h? m? x?

By now you probably realize that memorizing basic facts using flash cards does not help develop any notion for quantities. The only way to get a sense of the amount represented by a symbol is through experiences with that quantity of real objects and then to learn to associate the quantities you are experiencing with oral and written symbols.

Understanding Symbolization. Children who deal almost exclusively with symbols begin to feel that the symbols exist in and of themselves rather than as representations for something else. For example, Jackie's kindergarten teacher showed her a card with the numeral 8 written on it. When the teacher said, "Show me this many with the cubes,"

Jackie began to lay cubes out and then stopped and said, "I can't. The cubes won't make a circle." She was trying to make the numeral 8 with the cubes! It did not occur to Jackie to count out eight objects. For Jackie, 8 was *the symbol 8* and not what it represented. This is not to say that Jackie couldn't count to eight or that she couldn't get eight forks if her mother asked her to. What it does mean is that Jackie did not automatically associate the symbol 8 with the quantity of eight things.

Another child made the following mistake, even though he had objects to count. Tomás, a first grader, was working on the problem 1 + 8. He put out one object. Then, in attempting to count out eight more, he miscounted and actually put out seven. He then counted all the objects and arrived at eight. Without any hesitation at all, he wrote the answer to 1 + 8 as 8. He had gone through a ritual for determining answers to addition problems, got an answer, and wrote it down. The teacher then asked Tomás, "If you had one apple and you got eight more apples, would you have eight apples altogether?" He laughed and said, "Oh, no. I would have nine apples." Tomás knew something about number in the real world, but he did not use what he knew when solving a written problem. He did not even notice the inconsistency. The activity he was required to do at school using numerical symbols was unrelated in his mind to what he knew about numbers in the real world.

Gilbert, also in first grade, was extremely proud of the fact that he knew his "pluses." His mother had bought a set of flash cards and worked with him faithfully until he knew them all. One day the teacher was working with Gilbert. She put four cubes on the table and covered them with an overturned margarine tub. When asked how many cubes were under the tub, Gilbert said "Four." The teacher then showed him three more cubes and placed them under the margarine tub with the other four. Gilbert knew she had added three more cubes to the other four but could not see the cubes under the tub. The teacher asked him how many were under the tub now. He hesitated and then began to count on his fingers to figure it out. If she had shown Gilbert the flash card 4 + 3, he could have answered 7 without hesitation; however, when confronting the same situation with real objects, he saw no connection between the two.

It is essential that you not only help children develop number concepts through experiences with real things but that you also help them connect what they know to the symbols that represent these concepts. The connection between the experience and the symbol is critical. As long as children separate them and deal with them as two different things, they will have trouble using symbols appropriately. Our emphasis in schools must switch from encouraging children to write fast, automatic answers to written problems to helping children relate real experiences with the symbols that represent those experiences.

The activities in this chapter are designed to allow children to experience addition and subtraction as processes that occur in the real world and to help children develop a strong sense of quantity, number combinations, and number relationships. Symbols are introduced to the children as tools for recording their experiences with these processes and relationships.

Section I: Developing the Concept of Addition and Subtraction

The following activities are designed to help children understand the important ideas related to adding and subtracting. In this section there are three strands that are given special emphasis: the processes of addition and subtraction, number combinations, and determining sums and differences. One strand is not a prerequisite to the others, nor does one strand need to be mastered before another is experienced. Your lessons will sometimes emphasize one strand more than others and other times will include activities from two or all three strands. It is the integration of the ideas in all three strands that allows children to develop a strong sense of number and a clear understanding of addition and subtraction.

Because it is essential that children understand concepts before they are asked to deal with symbols representing those concepts, no written symbols are used at this stage.

The Processes of Addition and Subtraction: Teacher-Directed Activities

For children to clearly understand addition and subtraction, they must see the connection between these processes and the world they live in. Introduce your children to addition and subtraction by telling them stories, which they should act out. Begin by telling stories to the whole class at the beginning of your math period or at any time that you have available five to seven minutes. You do not have to worry about the readiness levels of any particular children. You will help them become ready by giving them these acting-out experiences in much the same way you prepare children to read by reading to them.

After they have had experiences using real things to act out the stories, you can have them use Unifix cubes to represent the people, animals, or objects in the stories. Using Unifix cubes not only allows a wide range of possibilities for stories but it also allows every child to be involved in the action of every story.

Your first goal is to familiarize the children with looking at groups of people or objects, seeing an action take place (joining or separating groups), and then determining how many objects or people are in the group after the action. As the focus is on the action (i.e., the process rather than the answer), it is important to include both addition and subtraction in random order from the beginning.

Your second goal is to acquaint your children with the specialized language patterns used to describe the actions of adding and subtracting such as *altogether, total, minus,* etc. Think of this goal as language development, in which the emphasis is on mathematical terms.

One of the problems that occurs when children are taught new vocabulary is that they begin to listen just for the particular words and not for the sense of the story as a whole. You can prevent this by giving them stories in which no mathematical clues are present right along with stories in which the mathematical terms are used. This way, the children will pay attention to the whole story and take note of the processes involved rather than just listening for the word clues. The following are examples of both types of stories.

Using mathematical terms: Linda stacked three books on the table. She *added* one more book to the stack. How many books are in the stack *altogether*? The words *added* and *altogether* are word clues particularly associated with addition.

Using no mathematical terms: Linda stacked three books on the table. She put another book on the table. How many books are on the table now? The process of addition can be extracted from the story, even though no mathematical terms are used.

Keeping the children thinking (looking for the sense of things) is more important than recognizing particular word clues. Ideally, both can be developed together.

Always keep in mind that the acting out of the addition and subtraction stories is to be a learning process and not a testing situation. So that no individuals are put on the spot, always have the group direct the children who are acting out the stories.

After an initial emphasis on acting out stories, spend five to seven minutes two or three times a week throughout the year to keep children in tune with the notion that addition and subtraction are part of the real world.

ACTING OUT ADDITION AND SUBTRACTION STORIES—Using Real Things*

Materials: Various objects readily available in the room

Tell addition and subtraction stories, and have the children act them out, using materials readily available in the room (such as books, chairs, pencils, etc.). These stories can be done with the whole class or with small groups. When working with the whole class, be sure you keep a checklist of the children who have had turns so that everyone will get a chance to participate. Present a mixture of addition and subtraction stories, some using mathematical terms and some not using them.

For example: Alice put four rulers on the table. Jim put two more rulers on the table. How many rulers are on the table? (Addition)

*Based on MATHEMATICS *THEIR* WAY, p. 204.

ADDITIONAL EXAMPLES:

Janann, Jack, Brad, and Lea are working at the table. Jack and Lea are excused to go outside. How many are still working? (Subtraction)

Susan, Laurie, and Todd are standing by the window. Tom and Kathy walked over to stand by them. How many children are standing by the window? (Addition)

Bob, Katy, Barbara, Cathy, and Bobbie were sitting in a circle on the rug. Barbara and Bobbie left the circle. How many children *remained* on the rug? (Subtraction)

Norma had eight crayons in her box. She took out the red one and the yellow one. How many crayons are *left* in the box? (Subtraction)

There are six children standing in a line by the door. There are four children standing in a line by the water fountain. How many children are there in both lines? (Addition)

There were eight pencils in the pencil can. John took two of them out. How many pencils are in the can now? (Subtraction)

Alex lined up six word cards on the chalk tray. Tillie took away four of them. How many cards are *left*? (Subtraction)

LuAnne poured three cups of water into the tub. Annie *added* three more cups of water. Cheryl *added* two more cups of water. How many cups *in all* are in the tub? (Addition)

Marilyn stacked up three pieces of paper on the desk. Bonnie stacked up four pieces of paper. How many pieces of paper are there *altogether*? (Addition)

ACTING OUT ADDITION AND SUBTRACTION STORIES---Fantasies

Materials: None required

Tell addition and subtraction stories in which the children pretend to be a variety of animals, people, or things. Because the goal is to relate addition and subtraction to the real world, it may appear that imaginary stories do not do this. Perhaps, then, we should use the term *natural* instead of *real*. The use of fantasy can provide more opportunities for the children to experience addition and subtraction than would be possible to encounter in regular classroom activities. Present a mixture of addition and subtraction stories, some using mathematical terms and some not using them.

For example: Adam, Timmy, Garland, and Freda are candles on the birthday cake. Robert blew three of the candles out. How many are still burning? (Subtraction)

Cindy, Bill, Andy, and Casey are clowns at the circus. Judy, Steve, and Jeff are more clowns who come to perform at the circus, too. How many clowns in all are in the circus? (Addition)

There were nine bees (Shirley, Carolyn, Linda, Eleanore, Fred, Chuck, Susie, Sally, and Mike) buzzing around the flowers. Three of the bees (Susie, Sally, and Mike) flew back to the beehive. How many bees are still flying around the flowers? (Subtraction)

James, Cathy, and David are boats sailing on the lake. Jimmy and Randy are rowboats on the lake. How many boats are on the lake? (Addition)

Sissy, Mark, and Nicky are birds sitting in a tree. Mark and Sissy flew away. How many birds are *left* in the tree? (Subtraction)

ACTING OUT ADDITION
AND SUBTRACTION STORIES—Using Unifix Cubes*

Materials: Unifix cubes • Counting boards—see p. 210 for directions for making • Working space papers—see p. 210 for directions for making

Tell stories and have the children act them out, using the cubes to represent the people, objects, or animals in the stories. Use either the working space papers to represent any setting you wish, or use a set of counting boards (cards on which are drawn settings such as trees, oceans, corrals, etc.), and make up stories for the setting being used. Have the Unifix cubes separated into the various colors so the children can easily find the colors they need.

For example:

Today we are going to make up stories using the corrals. What animals could be in our corrals?

Horses.
Cows.
Pigs.
There could be cowboys in there, too.

In this story there are five horses in the corral. Three of them are taken out to the pasture. How many horses are left in the corral?

Clear your boards so we can tell a different story.

There were three pigs in the corral. The farmer went to the market and bought three more. How many pigs altogether?

Other examples:

Counting board

There were seven apples on the tree. The farmer came and picked five of the apples. How many apples are on the tree? (Subtraction)

There were four ladybugs and three ants in the grass. How many were there *altogether*? (Addition)

Counting board

*Based on MATHEMATICS *THEIR* WAY, p. 206.

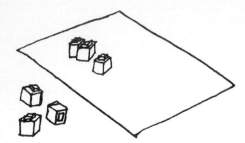

Working space paper

There are six cars in the parking lot at Palace Market. Greg, Troy, and Vesta drove their cars away. How many cars are *left*? (Subtraction)

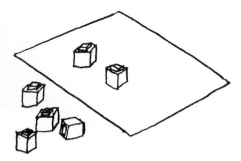

Working space paper

We gave our rat, Tippy, six kibbles. He ate four of them. How many kibbles are *left*? (Subtraction)

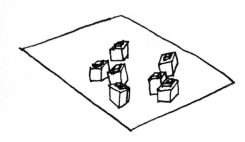

Working space paper

Four children were playing in the sandbox. Three more children came to play with them. How many children are in the sandbox? (Addition)

ADDITION AND SUBTRACTION STORIES——Extending the Language

Some mathematical terms appear to be difficult for children but actually are not. For example, the word *remains* may seem difficult for children to learn, but it is only another way of describing "what's left." That is not a difficult idea for children to grasp, so learning another word to describe the situation is no real problem.

On the other hand, there are situations that can be described by simple-sounding language but cause children some problems conceptually. Two of these situations involve missing addends and comparative subtraction. You will want to provide children opportunities to work with these types of stories, but you must be sensitive to the fact that they are not as simple as the examples given previously. Watch carefully to see how the children use the materials to work the following types of problems, as that will be a clue to the level of their thinking.

Missing Addend

Using real objects: 1. Rosalie is supposed to make sure there are six chairs at the round table. There are already three chairs at the table. How many more chairs does Rosalie need to get?

2. Deane checked out six books from the library. Four of the books were about dogs, and the rest were about horses. How many books were about horses?

Using Unifix cubes to represent objects: 1. There were seven pieces of fruit in the bowl this morning. Now only four are left. How many were taken?

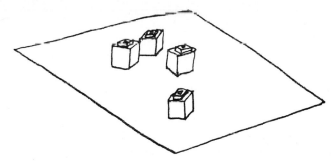

2. Norma has seven dogs at her house. Two of the dogs are grownups. How many dogs are puppies?

Comparative Subtraction

Using real objects: Linda picked out eight books from the classroom library. Shirley picked out five. How many more did Linda pick out than Shirley?

It is natural for many children to think that Linda picked out eight more books than Shirley. Help them focus on the meaning of that question by actually lining up the books. Put your arm across the books at the point where both are the same and tell the children, "These are the same." With your other hand point to the additional books and say, "Shirley has three more books than Linda."

Using Unifix cubes to represent the objects: 1. There are five horses and three cows in the field. How many *more* horses are there than cows?

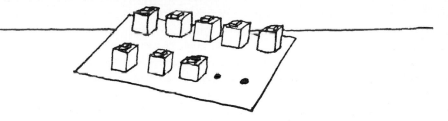

2. There was a freight train seven cars long carrying freight to the city. A passenger train with five cars was going to the city, too. How much *longer* is the freight train than the passenger train?

Help the children line up Unifix cubes in the same way the books were lined up so they can find where the two groups line up the same, and thus they can determine how many more are in one line than the other. Repeat these types of questions many times until the children are able to determine the answers easily.

Number Combinations: Teacher-Directed Activities

The following activities are designed to provide children many opportunities to learn the number combinations that make up any number from four to ten. Do the hiding assessment on page 127 of "Analyzing and Assessing Your Children's Needs" to determine what number or range of numbers each child needs to work with. Group children according to the assessment.

Using the number the group needs to work on, do three or four of the activities during a twenty-minute instructional period. It is not important which activity is used but rather that the children get a variety of experiences. Although the activities will vary, the combinations for the number being worked with will appear again and again.

The amount of time that any particular number will need to be worked with will vary according to the needs of individual children. A rule of thumb is that children should be confident with the combinations for four, five, and six and should have had experiences with the symbolization for these combinations before working with the activities in this section using the numbers seven, eight, nine, and ten.

Many children will benefit from working with one particular number for a week or so. At the end of that time, they should begin exploring the number that is one less or one more than the number with which they have been working. If they practice with one number too long, they tend to stop thinking and give automatic responses. Sometimes these programmed responses are given even when a different number is introduced! (For example, some children who have been working with combinations for five automatically say two and three when working with six instead of two and four.) Experiencing a new number even before the one worked with previously has been mastered encourages the discovery of number relationships and keeps the children's thinking alert.

Some children will need several weeks of working with four, five, and six. They should spend a few days with one number and move to a different one for a time. This could mean a week working with five, then to four, then to six, back to four, etc.

A few children will need just a few days focusing on each number and will then be ready for the symbolization of the combinations. (See "Analyzing and Assessing Your Children's Needs" for help in determining which children need which kinds of experiences.)

Do not rush your children to work with the numbers from seven to ten until they are able to deal comfortably and confidently with the numbers to six. Since the larger numbers are composed of the smaller numbers, success with the larger numbers depends a great deal on a firm knowledge of the relationships contained in the numbers six and under. Children who have a strong understanding of numbers to six and have learned to work easily with the symbols for those numbers have a better chance of understanding the connection between symbols and the real world. This understanding will carry over to their work with larger numbers.

Once children are able to deal comfortably and confidently with numbers to six, they are ready to work with seven, eight, nine, and ten. The nature of the activities changes when working with these larger numbers. Groups containing more than four or five objects are not easily recognizable at a glance unless they are organized in some fashion (or unless a person has been trained to organize them mentally). The following activities that use "number shapes" assume an important role for these larger numbers. Recommendations for changes in the other activities when larger numbers are used are noted in the descriptions of the activities when appropriate.

Many of the experiences with the larger numbers can be of a problem-solving nature, and the children will be able to move easily to using symbols to record their experiences.

Note: While you are encouraged to allow your children ample time to internalize the combinations for numbers to six before working with the combinations for the larger numbers, you should continue to give them other kinds of experiences with larger numbers (especially counting experiences such as determining boys and girls present, days until Christmas, and days of school).

SNAP IT

Materials: Unifix cubes

A group of children who need to work with the same number should be seated in a circle on the floor or at a table. They put together whatever number of cubes they are to work with for this period. On the signal "Snap," they all break their trains into two parts and hold them behind their backs. (They may also choose to break no cubes off and have all the cubes in one hand and none in the other.) Each child takes a turn (in order, going around the circle), showing what is in one hand and then the other while the other children say the combination shown. (The child showing the cubes should not say the combination. This will force the other children to look at the cubes to determine the combination formed.)

For example: All these children are working with five cubes.

Three and two.

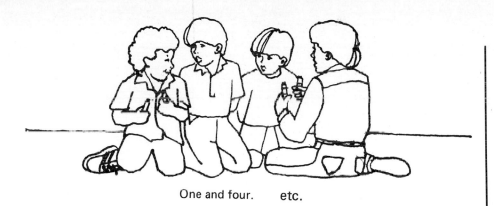

One and four.   etc.

When all the children have had a turn, they put the two parts back together; on the signal "Snap," they break their cubes into different combinations and go around the circle again. Repeat several times. The same combinations will appear again and again. Through this repetition the children will learn the combinations.

Extension: When the children are able to say the combinations for a particular number with little or no hesitation, have the child showing the cubes keep one hand behind his or her back while the other children predict how many cubes she or he is holding. The child can then show the cubes, and the other children can check their predictions.

Working with numbers above six:

When working with numbers above six, it is difficult to tell at a glance how many cubes are in some of the combinations. Instead of having each child form the combination of her or his choosing, you call out the number of cubes you want the group to break off their trains.

For example: *We are going to work with eight today. Snap together eight cubes. Now break off two.*

At your signal "How many?" the children say the combination formed. (Do not ask *how many* until it appears each child has determined the number of cubes in the combination.)

How many?   Six and two.

Ask the children to snap the cubes back together, and then ask them to break off a different number of cubes.

Break off five. How many?   Three and five.

During some lessons, have the children break off various numbers of cubes in random order. Other times have them break them off in order. For example, say, "Break off one. How many? Break off two. How many? Break off three. How many?"

THE TUB GAME

Materials: Unifix cubes • Margarine tubs (1 per child)

A small group of children who need to work with the same number should be seated in a circle on the floor or at a table. They each take the number of cubes being worked with for the day and place some under their overturned tubs and some on the tubs. (They may also choose to put no cubes on or under their tubs.) The children each take a turn showing what is on and then under their tubs while the other children say the combinations shown. When everyone has had a turn, they all change the arrangement of cubes on and under their tubs, and the activity is repeated.

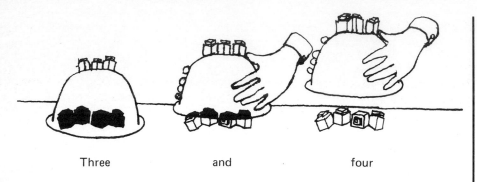

Three   and   four

Extension: When the children can say the particular combinations formed when using the tubs with little or no hesitation, have them predict what will be under the tubs *before* they are lifted. Then the tubs can be lifted and the children can check.

Working with numbers above six:

When working with numbers above six, it is difficult to tell at a glance how many cubes are in some of the combinations. Instead of having each child form the combination of his or her choosing, you call out the number you want the children to put under their tubs.

> For example: *Get seven cubes. Put four cubes under your tub. Put the rest on your tub.*

On your signal, "How many?" the children say the number of cubes on their tubs and then the number of cubes under their tubs. Repeat several times, having them form a variety of combinations with their cubes. During most of the lessons, say the number of cubes they are to place on their tubs in random order. Other days, present the number of cubes in sequential order (starting with one, then two, etc.).

THE WALL GAME

Materials: Unifix cubes • Working space paper (non-dotted side up—see p. 210)

Ask the children to line up whatever number of cubes they are working with for the day so that the line is pointing at their stomach (vertical line). They use their hands to make a "wall." You call out whatever number you want them to wall off.

For example:

Wall off four.

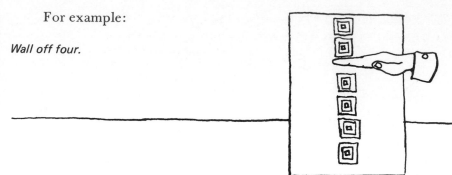

The children make a "wall" so that four cubes are close to their bodies.

Peek over the wall.

The children look over their hands and say the number of cubes behind their hands and then the number of cubes close to their body. In this example they would say, "Two and four." Continue to call out various numbers to be walled off, and they will continue to say the combinations formed.

THE CAVE GAME*

Materials: Unifix cubes • Working space papers

Ask the children to line up whatever number of cubes they are working with for the day in a horizontal line. (You can ask them to line up the cubes the way their arms go when they stretch them out to the side.) They place their right hands on the working space papers to form a "cave." (Have an outline of a right hand displayed in the room for those who need help in finding their right hands.)

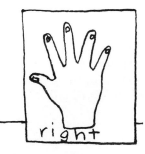

right

*Based on MATHEMATICS THEIR WAY, p. 192.

Tell them the number of cubes you want them to hide in the caves. On the cue "How many?" they tell the number of cubes outside the caves and then the number of cubes hidden inside the caves. On the cue "Check," they lift the caves and again say the combinations formed.

Hide three.

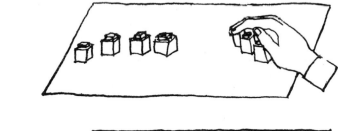

How many?

Four and three.

Check.

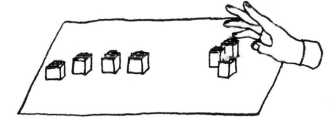

Four and three.

Continue to call out various numbers to hide, and they will continue to read the combinations formed.

FINGER COMBINATIONS

Materials: Unifix cubes

The children are seated in a circle on the floor or at a table. They take the number of cubes being worked with for the day and place some on the fingers of one hand and some on the fingers of the other hand. Each child takes a turn (in order, going around the circle), showing first one hand and then the other while the other children say the combinations shown. Repeat several times.

For example, if the children are working with six:

Three and three.   Two and four.   Three and three.

Allow the children to experience this activity with numbers above six, but let them discover on their own which combinations will fit on their fingers and which will not.

COUNTING BOARDS—Making Up Stories

Materials: Identical counting boards (1 per child)—see p. 210 for directions for making • Unifix cubes (sorted by colors)

Give each child one counting board. The following examples show the use of the oceans, but any of the counting board settings would work. Have the children take the number of cubes being worked with for the day and decide what they want them to represent. They take turns telling a story to go with the cubes.

For example:

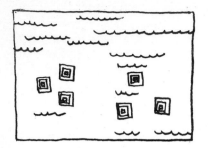

Three fish were swimming. Three more came. That makes six fish swimming.

Six ships were sailing on the water. Two sank and then there were four left.

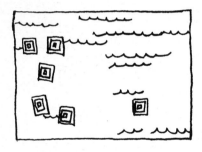

Five fish were in the water. One more fish came. Then there were six fish.

Six kids were swimming in the ocean. Four kids got cold and went home. Only two kids stayed longer.

NUMBER SHAPES

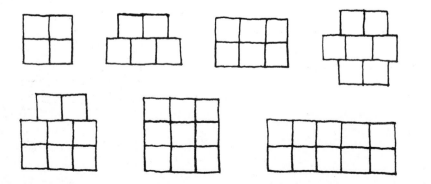

The number shapes are distinctive arrangements of squares for each number from four through ten which can be filled with Unifix cubes in various ways to show number combinations. The shapes have many helpful features:

> The arrangement of the squares in the shapes aid in the recognition of larger numbers. (Numbers above four or five are not easy to recognize without counting unless they are organized in some way.)

> The shapes provide specific images for children to picture in their minds, which will help when they are learning to visualize objects and the actual objects are not present.

Children who know that two plus three equals five do not automatically know that five take away three equals two, because they do not see the relationship between those two pieces of information. They seem to think about addition and subtraction as two totally different experiences. The shapes, however, can be used for both addition and subtraction, giving the children the opportunity to discover that the smaller numbers contained in any particular number are the same numbers that appear when they are taking that number apart. In most situations in which a child is subtracting, the original number being worked with disappears once he or she has taken a quantity away. This seems to be part of the reason young children have trouble relating the addition facts with the subtraction facts.

For example:

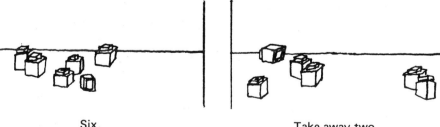

Six.   Take away two.

The six has disappeared and can't be referred to again unless the child is able to reverse her or his thinking and imagine the cubes back together again. This ability to reverse thinking does not develop for most children until age seven or later. Because they've lost track of what they started with (the whole group), many children will say or write $4 - 2 = 4$ rather than $6 - 4 = 2$. When the same problem is done on the shape, the shape serves not only as a reminder of how many were started with but also, because the empty spaces are still there, presents an image of the 6 as it relates to the 4 and the 2.

Working with the Number Shapes

Materials: Number shapes—see p. 212 for directions for making • Unifix cubes

Give each child in the group you are working with the same size number shape. Direct the children to do a series of actions with the shapes. It is important that you use the shapes in a variety of ways from the first day the children work with them so they will be able to deal with them flexibly.

The following examples use the "six" shape as the number being explored for the day. The same activities can be used for any of the shapes.

Following directions:

Here are some examples of the kind of directions you can give the children. These include addition, subtraction, and determining how many more. Addition problems are done with two colors to highlight the two sets. Subtraction problems are done with one color to emphasize the whole group. Giving children a variety of directions during one lesson will keep them flexible and alert.

Put three red and three blue cubes on your shape. That shows three and three are six.

Clear your shape. Now put four red cubes and two blue cubes on your shape. Who can tell me what that shows?

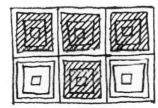

Four plus two equals six.

Note: Children may describe the action in a variety of ways such as "four and two is six" or "four and two more is six altogether." Accept a number of expressions

Fill the top row with red cubes. Fill the bottom row with blue cubes. What do you have?

Three plus three equals six.

Put one red cube on your shape. Fill the rest of the shape with blue cubes. What do you have?

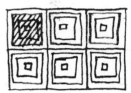

One plus five equals six.

Put two red and four blue cubes on your shape. What do you have?

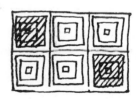

Two plus four equals six.

Can you find another way to arrange those cubes on your shape? How did you arrange them?

I put two on the bottom corners and the rest red.

Can you find another way? How did you arrange them this time?

I put two red ones in the middle.

Fill up your shape with blue cubes. Take one off. Who can tell what we did?

Six minus one is five.

Put two cubes on your shape. How many more do you need to fill up the shape?

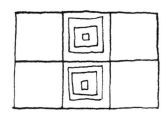

Four.

Yes. Two and four more makes six.

Describing an action with number sentences:

You and/or the children can perform a variety of actions while the others describe what you did. The following are examples of what could be done.

The teacher puts six cubes on the shape and takes four off.

Tell me with numbers what I did.

Six minus four equals two. or
Six take away four leaves two.

The teacher puts four red cubes on the shape and then puts on two blue cubes.

Tell me with numbers what I did.

Four and two makes six altogether.

Using dice to create number combinations:

The children will be highly motivated to create a variety of combinations with the number shapes if they can use dice to determine the combinations they are to make. You will need three sets of dice: 0-4 dice, 1-6 dice, and 4-9 dice. The 0-4 dice can be used for the four and five number shapes, 0-4 and 1-6 dice can be used for the six, seven, and eight number shapes; and 0-4, 1-6, 4-9 dice can be used for the nine and ten number shapes. (See notes on making, color coding, and using dice on p. 215.)

Addition: Have the children take turns rolling a die to see what number of cubes they are to place on their number shapes. They place that number on the shape and fill in the remaining spaces with cubes of a different color; they then describe the combinations formed.

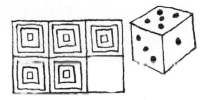

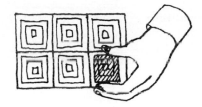

I rolled five.

I need one more.

Five plus one is six.

Subtraction: Have the children fill the number shapes with cubes of one color. Roll the dice to see how many they are to take away. Describe the action with numbers.

I rolled four.

Six minus four equals two.

Note: Encourage the children to put cubes on and take cubes off in groups. Do not force or require them to do this, but focus their attention on the technique so those who are ready can pick up on it and begin to use it.

For example, you may say something like the following:

When I take four away from seven, I don't have to count by ones and say one, two, three, four. I can look at the cubes and see that two and two are four, so I can take four away all at once.

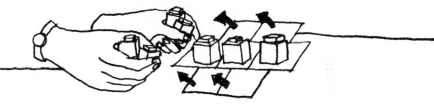

Many children will already be doing this on their own; others will need you to point out the "easy" way. You will probably need to repeat this kind of discussion every so often, as children will not necessarily pick up on it right away.

NUMBER TRAIN OUTLINES

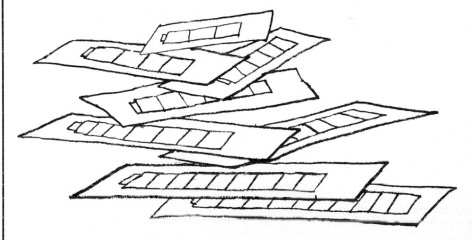

The number train outlines are drawn the same size as the Unifix cubes so that when the cubes are snapped together, they fit into the outlines exactly. The numbers represented by the train outlines are not instantly recognizable as they are in the number shape arrangements. They do, however, provide children an alternative way to experience the number facts and number relationships. It is important that the children use the outlines in a variety of ways from the first time they are exposed to them so they learn to deal with them flexibly.

The following examples use the "seven" train outline as the number being worked with for the day. The same activities could be used for all the outlines.

Working with the Number Train Outlines

Materials: Unifix cubes (2 colors) • 1 number train outline per child (appropriate number for the day)—see p. 212 for directions for making

In this example, all the children in the group you are working with should have a "seven" train outline. Note that two colors are used when the focus is on addition. One color is used when you want the children to focus on the whole group before subtracting.

Following directions:

Give the children a variety of directions to follow. Include addition, subtraction, and comparing numbers. The following are examples of directions you can give.

Using the green and yellow cubes, snap together enough to fill in your number train outline. How many did you need?   Seven.

Hold up one of the children's trains.

How many green cubes did Jane use?   Four.

How many yellow cubes did Jane use?   Three.

Yes, four and three is seven.

Hold up other children's trains and have the group read the combinations contained in each train. Keep the color they are to read first consistent—in this case, first green and then yellow.

Now fill your outline with green cubes. Take away four cubes. How many are left?

Three.

Yes, seven take away four is three.

Repeat the activity, asking the children to take away a variety of numbers from seven and saying the equation.

Ask the children to break off cubes in a pattern.

Break one cube off. How many left?
Break two cubes off. How many left?
Break three cubes off. How many left?

Can you guess how many I'm going to say next? How many do you think will be left?

Put three cubes on your outline. How many more cubes do you need to fill in the outline?

Repeat, varying the numbers.

Hold up a train seven cubes long. Show the children that the train can be described in two ways.

How many yellow cubes? How many green cubes? Yes, three plus four equals seven. We can read this train another way, too. We can also say one plus four plus two.

Have the children take turns holding up their number trains so the others can read them by color and by arrangement of cubes.

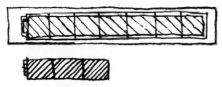

Fill your outline with green cubes. Make a yellow train three cubes long and place it next to your green train. How many more cubes do you need to make your yellow train as long as your green train?

Repeat, varying the length of the yellow train.

Show me one plus six with your cubes. Can you make one plus six a different way? Another different way?

Repeat, varying the combinations.

Describing an action with number sentences:

You and/or the children can perform a variety of actions while the others describe what you did. The following are examples of what could be done. The teacher snaps seven cubes together and lays them on the outline. He breaks off four.

Tell me with numbers what I did.

Seven minus four equals three, or seven take away four leaves three.

The teacher puts two blue cubes in the "seven" outline. He then puts five red cubes in the outline.

Tell me with numbers what I did.   Two plus five equals seven.

Ask the children to demonstrate for each other so the rest can describe what happened.

Using dice to create number combinations:

Use 0–4 dice for 4 and 5 trains, 0–4 or 1–6 dice for 6, 7, and 8 trains, 0–4 or 1–6 or 4–9 dice for 9 and 10 trains.

Addition: Have the children take turns rolling the dice to see what number of cubes they are to place on the outline. They place the number of cubes according to what they rolled on the dice. Then they fill in the remaining space with cubes of a different color and describe the combination formed.

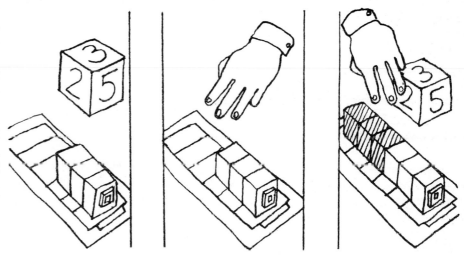

I rolled three.   I need three more.   Three and three is six.

Subtraction: Have the children fill in their number train outlines with cubes of one color. Have one child roll the dice to see how many they are to take away. Then have the children describe the action with numbers.

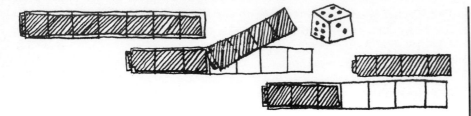

We need to break off four. Seven minus four equals three.

Encourage the children to break off the number rolled as a unit rather than one by one.

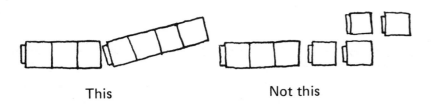

This   Not this

GRAB BAG SUBTRACTION

Materials: Unifix cubes • Paper sack

Put the number of cubes being worked with for the day into a paper sack. Have the children take turns reaching into the bag and taking some out. The child whose turn it is shows how many he or she took out, and the rest of the group tells how many they think are left in the sack. Then the cubes are dumped from the bag and the guess is checked.

For example, if the number for the day is eight:

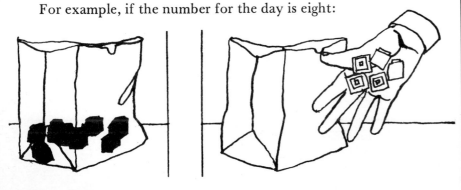

Eight.   Take away five.

How many do you think are still in the bag?

Three. Maybe four.

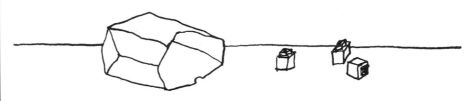

Three are left.

Let's say what happened.

Eight take away five is three.

BULLDOZER

Materials: Unifix cubes • Working space papers

Have each child put out the number of cubes being worked with for the day on his or her working space paper. At your direction, they push off the number of cubes you say and then tell how many are left.

For example:

Put on six.   Push off four.   How many are left?

Two.

Variation: Have each child take a turn pushing off a number of cubes while the rest of the children tell how many were removed and how many are left.

For example, one child performs the action:

The other children say:

Eight.   Push off two.   Leaves six.

Number Combinations: Independent Activities

The activities in this section are of an open-ended, problem-solving nature. The full potential of the activities can only be realized if the children have many opportunities to work with them.

The independent activities are so defined because they can be done by children working alone or with partners. The teacher does not need to be directly involved. These activities are useful, then, to engage children in meaningful tasks while you are doing teacher-directed activities with small groups. You will find, however, that you can learn much about your children's levels of skill and understanding as well as their approaches to open-ended situations if you also assign these tasks during periods when you can observe them working.

Children of many different levels can do these activities working side by side. Their individual needs can be met by assigning different numbers to children who have different needs. For example, some children may be working with combinations of five while other children are working with combinations of eight. (See "Analyzing and Assessing Your Children's Needs" for help in determining which number is appropriate.)

NUMBER TRAINS*

Materials: Uniflx cubes (2 colors)

Using two colors, the children snap together as many different arrangements of cubes as they can for the particular number they are to work with for the day.

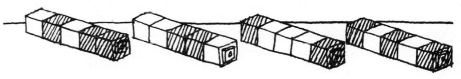

When you observe your children, you will see them approaching this task in many different ways. Some will immediately get involved, making lots of arrangements. Others will have a difficult time getting started. There will be some who give up after finding just three or four ways.

Your role as teacher in these settings is very important but very sensitive. You must find ways to help children move ahead in their own ways and to accomplish all they can without imposing on them your own way of doing the task.

For example: Matthew has made three trains. He is just sitting there, and, when the teacher walks by, he says, "I can't think of any more."

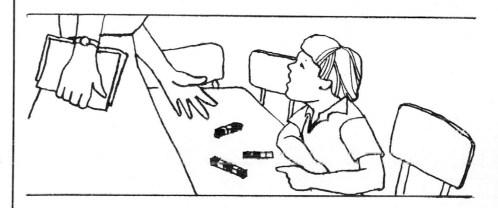

*Based on MATHEMATICS *THEIR* WAY, p. 172.

The teacher needs to help Matthew realize there are many more ways. "Can you make a train in which the blue ones are not touching? Can you make a train with three green? Can you make a train with just one green that looks different from this one? Can you make another one? I'll be back in a few minutes to see how many you have made."

Joseph is busy making trains, but he repeats the same two or three arrangements over and over. The teacher helps him focus on the trains he has made that are alike. "Can you find a train that looks just like this one? Are there more? Now find one that looks different from these. Can you make another one that looks different from all the others? What if you used one yellow? Three yellow?

Concha has found numerous ways of making trains.

The teacher can see by the way she has laid out her trains that she has approached the task in an organized fashion. Concha does not really need much input from the teacher. The teacher poses a couple of questions to see if Concha can verbalize any of the thinking she has been doing. "Tell me about your work. Do you think you've found all the ways to make six? How do you know? Did you find more or less ways when you worked with five?

Margie is very involved. She has made many arrangements and is searching for more. The teacher does not interrupt Margie while she is so intent. Even praise would not be helpful at this point. Margie is not working to please the teacher; she is working to please herself. When the level of Margie's involvement decreases, the teacher discusses Margie's work with her. "How many ways did you find? How many ways using one red? Two reds? I see some arrangements of 3 + 2 + 1. Can you find them? Tell me some other combinations you made. How many greens in this train? If I took them away, how many would be left?

Becky has found numerous combinations but has not organized them in any way. Her teacher remembers that Becky had a hard time thinking of more than three or four ways for the first few days that she was assigned this task. He is pleased to see how freely she is working now. The teacher knows that Becky must find her own way of organizing her work and he does not want to impose on her his personal way. However, he will ask Becky some questions that will focus her attention on ways that her trains are related: "Can you find all the ways with one red and the rest white? With two reds? Look, you made two that are opposites—this one is two red and three white, and this one looks the same, except it has two white and three red. Can you find any more that are opposites?" If any of his questions seem not to make sense to Becky, he simply moves on to some other questions. The teacher will watch Becky over the next few days to see if she is beginning to organize her trains in any way.

Extension: Making Records*

Materials: Worksheets (see black-line masters 41–48) • Unifix cubes • Crayons • Dice (optional)

a. The children can make records of their work by coloring in number train outlines to match the Unifix trains they have created.

b. The children roll dice to indicate the number of cubes to add or subtract.

Addition: The children can roll dice to determine the number of cubes of one color to place in number train outlines reproduced on worksheets. After filling their outlines with the number rolled of one color of cubes, they are to fill the rest of their outlines with a second color. They then color the outlines to match the trains they built.

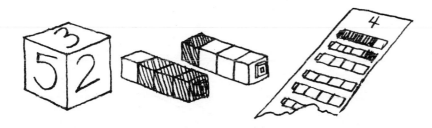

Subtraction: To subtract, the children can fill their number train outlines with cubes of one color. They then roll the dice to determine how many to take away. After taking the cubes away, they color in the outlines to show how many are left.

*Based on MATHEMATICS *THEIR* WAY, pp. 178, 239.

Note: While the worksheets do not have to be cut apart, there are some advantages to doing this. If the worksheets are cut apart, the children will be able to complete as many as they can rather than a predetermined number assigned by the teacher. Thus, some children will be able to complete many more than you may have assigned, and other children will not be frustrated with an assignment they can't finish in the allotted time. These recordings can be stapled into a little booklet. (If you want the children to be assigned a specific number of trains to record, you can have them complete one worksheet.)

NUMBER ARRANGEMENTS

Materials: Unifix cubes (1 color)

The children create a variety of arrangements of cubes for the particular number being worked with for the day. Generally, the cubes will be loose, but it is all right to use some cubes snapped together as part of some arrangements.

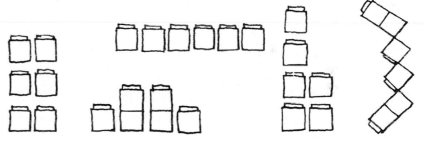

At those times when an adult is available to the children, he or she can ask the children to describe their arrangements with numbers. The arrangements can be described in a variety of ways:

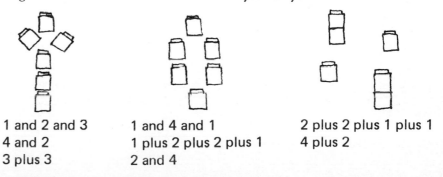

1 and 2 and 3
4 and 2
3 plus 3

1 and 4 and 1
1 plus 2 plus 2 plus 1
2 and 4

2 plus 2 plus 1 plus 1
4 plus 2

Variation: You can raise the level of difficulty for some children by posing some special challenges such as:

How many arrangements can you make with full sides touching?

This   Not this

With corners "kissing"?

Can you make a square with your number?

How many different rectangles can you make with the number you're working with today?

Note: When open-ended tasks are assigned as independent work, there are both advantages and disadvantages. An advantage is that when the assignment is to "do as many as you can," no child can finish early and need an additional task. Because you have set no limits, some children will accomplish much beyond what you may have assigned if you had established a specific amount to do. The disadvantage is, however, that other children will accomplish much less than you may have assigned, because they feel no pressure to complete a particular amount.

Give only those children who need a specific assignment a certain amount to do. This can be done by providing the child with a "workboard" to be filled in. This defines a specific amount to be accomplished.

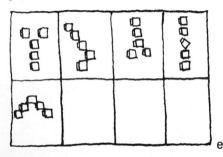

etc.

The workboard can be made from a 12 x 18 piece of paper folded into eight sections. The child is to put one arrangement in each section. Thus, she or he will have to complete eight arrangements in order to finish the task.

Extension: Making Recordings

Have the children make records of their work by gluing colored paper squares (to represent the cubes) to 6 x 9 pieces of paper.

(Squares from the Unifix Album of Gummed Stickers may also be used—see the list of commercially available materials on p. 217.)

Note: Be aware that making records can hamper a child who could be very creative and productive if not slowed down by the recording process. Be sure you provide children opportunities to explore the problem posed using just the materials with no requirements to record the work. Sometimes you can have the children make arrangements for a certain period of time and then just record a few of their arrangements.

NUMBER SHAPES

Materials: Unifix cubes (2 colors) • 8–10 number shapes for each child of the number to be worked with for the day—see p. 212 for directions for making

Have the children arrange cubes of two colors on the number shapes in a variety of ways.

Extension: Making Records

Materials: Unifix cubes • Worksheets 34–40 • Crayons

Have the children color in the appropriate worksheets to match their arrangements. The worksheets can be used in two ways: They can be cut apart and the children can do as many as they are able or the whole worksheet can be completed.

a. The children create their own arrangements of cubes.
b. The children use dice to determine the number of cubes to add or subtract.

Addition (using dice): The children roll the dice to determine the number of cubes of one color to place on the number shape. After placing the appropriate number of cubes of one color on the shape, they fill in the remaining spaces with cubes of another color. Then they color in the number shape on the worksheet to match their arrangement.

Subtraction (using dice): The children fill the number shape with cubes. They roll the dice to determine how many cubes to take off. After they have taken the appropriate number of cubes away, they color the number shape worksheet to match the results.

COMPLETE A FLOOR RACE—A Game for Partners

Materials: Game board (run off on construction paper or tagboard—black-line master 49 • Dice • Unifix cubes (2 colors)

The object of this game is to be the first to complete a Unifix cube floor. The floor is made of Unifix trains lying side by side. Each train is made of two colors. Various size floors can be made, depending on the number combinations you want the children to work with.

Combinations of 4   Combinations of 6   Combinations of 7

For example: The completed floor in this example is to be five cubes long. In the first stage of the game, the children take turns rolling dice to determine the length of the first part of each train. Each child rolls six times, putting out the number indicated. Numbers can be repeated.

Partner A   Partner B

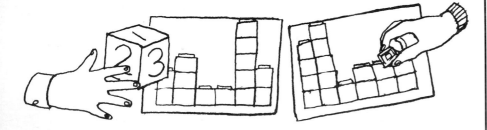

In the second stage of the game, the children take turns rolling the dice to complete their own floors. A different color cube is used.

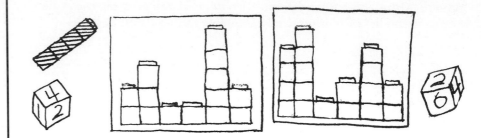

I rolled a four. I can put it with the one to make five.

I rolled a two. I don't have a three on my board, so I can't play this one.

The children continue to take turns rolling the dice until one player has completed a floor which in this case is

I still needed to roll a one and a four before my floor would be done.

Determining Sums and Differences: Teacher-Directed Activities

The following activities focus on the answers rather than on the processes of addition and subtraction or on the combinations that make up a number. Children focused entirely on the process of adding or subtracting often go through a counting ritual and write an answer without ever being conscious of what answer was arrived at. Children who have worked with the quantity of six may know the combinations for six. But it is one thing to know the parts that make up six in a setting where six is the focus and another thing to recall that four and two are six when presented those numbers mixed in with many others.

The following activities provide children opportunities to focus on addition and subtraction in a different way. In a sense, they are asked to apply what they have been learning in other lessons to these new situations. Resist the feeling that these are culminating activities or that they are more important than the others. These activities are valuable only in the context of the rich and varied experiences the children are having through the lessons from the other sections. The *variety* of approaches and experiences develop the concepts of addition and subtraction.

As long as the children are experiencing number in diverse and meaningful ways, it can be useful at times to provide them opportunities (as presented in this strand) to try and remember sums and differences, to practice visualization of quantities, and to practice finding answers efficiently when they can't remember them. For many children these kinds of activities are a means of discovering how much they already know about numbers.

Once every week or two, include some of the following activities in your lessons. Work with the numbers that the children have experienced previously. For example, if the children have been working with combinations of six and seven, work mostly with three, four, and five and a little with six and seven. When the children begin working with the combinations for eight and nine, you will want them to experience the following activities with numbers to seven.

Remember that these activities are to be used for teaching—*not* testing. Be very aware of this as you are working. Children should always have a way of finding out an answer they don't already know. If that opportunity is not present, the activity is a test of what is already known. Tests give us information about what has or has not been learned and therefore what to teach, but they do not themselves teach.

INSTANT RECOGNITION OF COMBINATIONS

Materials: A variety of recordings made by children when working with Unifix arrangements—see p. 98

Have the children tell as fast as they can how many squares they see in a variety of Unifix cube arrangement recordings. Begin using arrangements of five or less. Hold up one arrangement at a time and ask, "How many?"

For example:

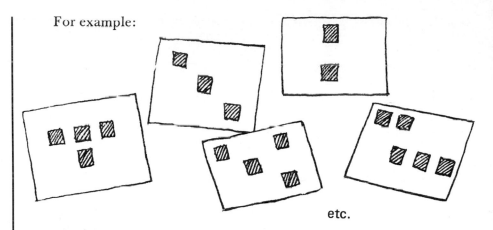

etc.

When the children can instantly recognize groups of five or less, include groups of more than five. In order to recognize groups larger than five quickly, the children will have to combine mentally the smaller groups they see contained in the larger numbers.

For example, hold up an arrangement.

Tell me fast. How many?

Seven.

*How did you know?**	'Cause there's four and three, and that's seven.
Who found out a different way?	I saw four and then counted the others . . . five, six, seven. So that's seven altogether.
Let's try another one.	
How many?	

Six.

How did you know?

I saw four on the top and two hanging down. That's six.

Who found out a different way?

I saw four in a line down the middle and one on that side and one on that side. That makes six.

How many in this one?

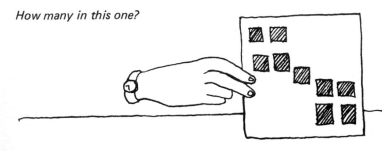

Nine.

How did you know?

'Cause four and four is eight and one more makes nine.

Note: Because it is not always easy for young children to verbalize their mental strategies, allowing them to frame each group with their hands helps them organize what they want to say. Before you begin asking the children to tell how they knew what number it was, model for them by saying such things as "I knew it was seven because I saw four on top and two there and one more." Allowing time for language development is very important. Experiences such as this are valuable organizers for clear thinking.

COUNTING TO FIND ANSWERS

When children do not know answers, they need to count to find out. Counting on and counting back can be useful skills for them to develop. Some children need to practice these skills before they will be able to use them to find answers. The following activities provide opportunities for them to become familiar with counting forward from any number and counting back from any number.

Counting On

Materials: Unifix cubes • Dice

The following activity is designed to give the children practice starting in the middle of the counting sequence and then counting on to find out how many altogether.

Tell the children to build a Unifix train of a certain length.

For example: Have them build a yellow train that is three cubes long. Then have a child roll a die to determine how many cubes are to be joined with the yellow train.

John rolled a four. Put four red cubes with the yellow cubes. Let's count to see how many. How many yellow?

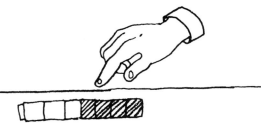

Yes, three, four, five, six, seven. There are seven altogether.

Have the children keep the original yellow train of cubes, and break off the red cubes. Ask another child to roll the die to determine how many to be added on this time. Repeat several times, beginning with the original train of three (or whatever number is appropriate).

Counting Back

Materials: Unifix cubes

Many children can say the counting sequence backward, starting from ten and ending with "blast off!" That does not necessarily mean they can start at any number and count backward, nor does it mean they can *use* the counting backward sequence appropriately.

When doing a problem such as 8 − 1, many children will take one cube away and then count the rest of the cubes to determine how many are left. They need to do this because they don't know the counting sequence backward. To know there are seven left, you need to know that 7 is the number that comes before 8. (To understand this, see how quickly you can say the letter that comes before R in the alphabet. Unless you have had lots of experience putting things in alphabetical order, you probably had to start saying the alphabet going forward from some point in order to find out what comes before R. It's not that you don't understand; it's simply that you haven't practiced or experienced the alphabet backward.) Learning to count backward is not the solution to subtraction for young children, but it can be a useful tool for them to have—especially when they need to subtract just one or two.

Spend a few minutes now and then having the children practice the counting sequence backward, starting with various numbers of cubes. Begin with small numbers (3 or 4), and make sure students are comfortable with this short sequence before working with the longer sequence. For example, tell the children:

Snap together four cubes. Let's break one cube off at a time and count backward to see how many are left.

Initially, with small numbers of cubes, the children will be able to look at their cubes and say the appropriate number. To see if they can say the sequence with no visual clues, have them say the sequence as they pretend to break cubes off an imaginary train.

As the children gain confidence counting backward from small numbers, begin to lengthen the sequence of numbers with which they are practicing.

Counting the Stacks

Materials: Unifix cubes

Give the children practice counting on and counting back by showing them a series of Unifix cube stacks one at a time and asking them to tell how many they see. Encourage them to tell how many as fast as they can, and discuss with them the ways they found out.

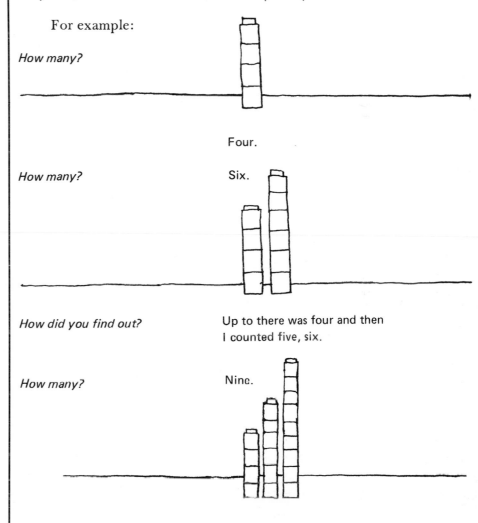

For example:

How many?   Four.

How many?   Six.

How did you find out?   Up to there was four and then I counted five, six.

How many?   Nine.

How did you find out?   I knew that was six. So that was seven, eight, nine.

How many?   Eight.

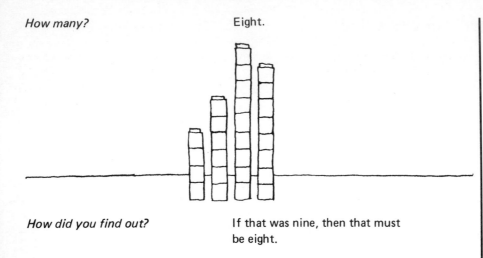

How did you find out?   If that was nine, then that must
                               be eight.

Note: Some children may know how to count on or count back but will not use the skill in this situation or when adding or subtracting. Do not force them to use it. Many children still need to count from one to be *sure* of what is there. They will use the counting on and counting back skills when they are ready, if they have been exposed to these techniques in settings such as this.

ADDITION TELL AND CHECK

In this activity, a variety of number combinations are presented, and the children are asked to tell the sums. The combinations are presented with real materials rather than written cards. These activities provide children an opportunity to try and remember what they have learned in other settings. For those children who don't remember addition facts, the activities provide the opportunity to count on and experience that number combination one more time.

Four steps are involved in the activity:

1. Noting the groups to be added.

2. Telling the sum, if the child knows it. (Some children know the combinations but think they are supposed to count as part of the routine of adding and subtracting. This will encourage those children to use what they know.)

3. Counting on to check. (It won't be necessary to check each time, as you want the children to know it is possible to know a combination without counting. However, if you check consistently, it will help those children who don't remember and confirm the facts for those who are just learning them.)

4. Repeating the equation. (If you repeat the whole equation each time rather than just the answer, it will be one more way for the equation to be imprinted in the children's minds.)

You can follow the steps outlined above using a variety of previously experienced activities.

Addition Tell and Check—Using Tubs

Materials: Margarine tubs • Unifix cubes

How many on top?

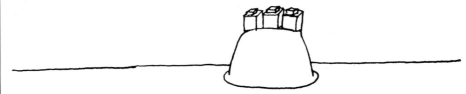

                               Three.

How many underneath?

                               Four.

What is three plus four?

                               Six.
                               Seven.
                               I don't know.

Let's check and see how many.

Three   four, five, six, seven

*Let's say the whole equation: three
plus four equals seven.*

Repeat, using a variety of numbers within the assessed range.

Addition Tell and Check—Number Trains

Materials: Unifix cubes (previously joined together)

How many blue?	Three.
How many yellow?	Three.
What is 3 and 3?	Six.

Let's count and check.   Three (saying the number of yellow cubes).

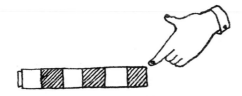

Four, five, six.

Say the whole equation.

Three, and three is six.

Repeat, using a variety of trains within the appropriate range.

SUBTRACTION TELL AND CHECK

Materials: Unifix cubes · Working space papers

Provide the children opportunities to work with a variety of subtraction equations.

Unifix Trains

Hold up a Unifix train.

How long is my train?	Five.

(Break some off.)

How many did I break off?	Three.
How many are left?	Two.
Say the whole equation.	Five take away three is two.

Repeat, using a variety of number trains.

Working Space Papers

*Put six cubes on your board. Take
four off. How many are left?*

Say the whole equation.

Two. Six minus four equals two.

Repeat, using a variety of numbers.

Grab Bag

Have the children count with you as you place the cubes in the bag.

We have eight cubes in the bag.

Ask a child to take some out.

We took away three. How many are left?

Repeat, starting with a different number of cubes each time.

DEVELOPING VISUAL IMAGES

You can help children toward the realization that they can "see the cubes in their heads" by exploring the following questions with them. Not all children will be ready for this, so treat it lightly. If it seems confusing to some children, do not ask these types of questions for a while. The problem for the children who are not ready to see the cubes in their heads is that they may begin to feel the other children are "magic." Their faith in their own ability to know can be damaged. The following activities should help the children feel more confident in their own power to know. If you look at the children's reactions (i.e., do they look puzzled? Are they waiting for others to answer? Are they saying incorrect numbers?), you will know whether or not these activities are appropriate.

Grab Bag

Materials: Unifix cubes • Paper bag

I put two cubes in. Now I am putting four cubes in. How many are in the bag now? Repeat, varying the numbers used.

Let's Pretend

Materials: Unifix cubes • Paper bag

Let's pretend I put five cubes in the bag, and then I put in three more cubes. How many cubes will be in the bag? Can you see the cubes in your head? Let's do it with the real cubes now and see if we were right. Repeat, varying the numbers used.

Counting Boards

Materials: Unifix cubes • Counting boards

Tell stories, and have the children predict what will happen and then check. For example:

Four ladybugs were crawling in the grass.

The children put four cubes on the grass counting board.

What will happen if two more ladybugs come? How many will there be?

The children tell how many they think there will be.

Let's check and see.

Use real cubes to allow children the opportunity to check their predictions.

SEEING RELATIONSHIPS BETWEEN COMBINATIONS

Materials: Unifix cubes • Margarine tubs

Knowing that three plus three equals six does not necessarily help a child know that three plus four equals seven. Presenting number combinations in a way that highlights the relationships between number combinations will help some children discover that they can use what they know to figure out what they don't know. Not all children will be ready for this, but it can be an interesting discussion for those who are ready to think about it.

How many on top?

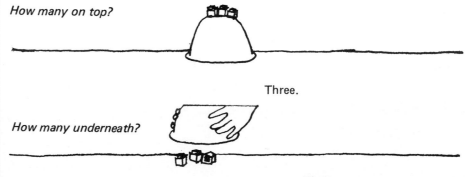

Three.

How many underneath?

Three.

How many altogether?　　　　　　　Six.

*Now I'm going to place a cube under
the tub. Now how many altogether?*　Seven.

How did you know?　　　　　'Cause you put one more in.

Now how many on top?

Three.

How many underneath?

Four.

Yes, three plus four equals seven.

Let's try another one.

How many on top?

Four.

How many underneath?

Four.

How many altogether?　　　　　Eight.

*Now I'm taking one away. How
many now?*　　　　　　　　Seven.

How did you know?　　　'Cause you took one out so now it's
　　　　　　　　　　　　　four plus three, and that makes seven.

PICK IT UP

Materials: Unifix cubes

Show the children five Unifix stacks (each made with a different color
and no stack more than five cubes high), and have the children build
stacks to match those you are showing. The object of this activity is
to find out what numbers it is possible to pick up using whatever stacks
you have made for this game. The rules are that you can pick up more
than one stack at a time, but you may not join any together and you
may not take any apart.

　For example:

*Build these stacks. Make them
exactly like mine.*

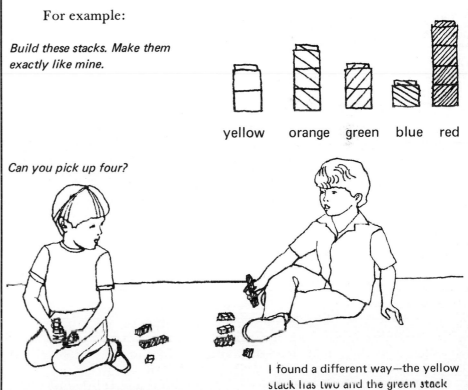

yellow　orange　green　blue　red

Can you pick up four?

I can. I picked up the red stack.

I found a different way—the yellow
stack has two and the green stack
has two. That makes four.

Can you pick up five?

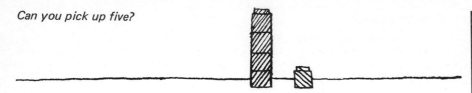

Yes. The red four and the blue one.

There's another way—the orange three
and the yellow two.

I used the orange three but I used the
green two.

This game can be played over and over; just have the children build different trains. Explore such things as:

What happens when you don't have a "one" stack?
What happens when all the stacks are the same number?
What happens when you use all even numbers? All odd numbers?
What's the largest number we can make with these stacks? with any stacks?
What's the smallest number we can make?

Section II: Connecting Symbols to the Concepts of Addition and Subtraction

There are three purposes for the activities in this section:

1. Helping children understand that symbols do not exist in and of themselves but are ways of recording experiences.
2. Providing children opportunities to learn to write addition and subtraction equations in the correct form.
3. Giving children practice using addition and subtraction to record a variety of experiences.

The children should be able to interpret the language of the addition and subtraction stories presented in Section I, before being introduced to the symbols. If they are able to act out the stories easily, then it is appropriate for you to begin modeling the equations for them as described in this section.

It is likely that you will introduce the symbols as labels for addition and subtraction stories before you use symbols to record the number combinations. The children can then be learning to use symbols to record equations at the same time they are learning number combinations *without* symbols. The children should know the combinations for four, five, and six before being asked to deal with the symbols for those combinations. (See "Hiding Assessment" on page 127.)

The independent activities in this section require that the children already know how to write addition and subtraction equations. They provide practice writing equations in a variety of situations.

Connecting Symbols to the Processes of Addition and Subtraction

MODELING THE RECORDING OF ADDITION AND SUBTRACTION

Once the children are able to act out the stories you tell (see description, page 81), begin modeling the writing of equations to describe the stories. They do not need to have had any previous experiences with plus, minus, or equal signs. The symbols will come to have meaning for the children when they watch you write the equations as the stories are being acted out.

It is essential that your students come to see these symbols as ways of writing down their experiences; therefore, after the equations have been written, read them back to the children in their natural language to encourage the connection of those symbols to the story being acted out. For example, if the children have acted out a story using pencils and the equation $4 + 2 = 6$ was elicited, read it back as "Four pencils and two pencils are six pencils." After many experiences, when the children are confident expressing an equation in terms of the situation from which it came, introduce them to another way of reading the equation: "Four plus two equals six." From then on, use the natural language and the formal mathematical way of reading the equation interchangeably. The following examples can be acted out by children or with Unifix cubes.

Addition:

Linda, Paul, Lupe, and Maria are in line to jump rope. How many are in line? . . . Yes, four.

(Write *4* on the chalkboard.)

Dick and Bill got in line to jump rope too. How many got in line? . . . Yes, two.

(Write *4 + 2*.)

How many children are playing jump rope? . . . Yes, six.

(Write *4 + 2 = 6*.)

This is one way to write about the story we just acted out. It says, "Four and two is six." Four children and two children are six children altogether.

Subtraction:

There were five cupcakes on the plate. How many cupcakes?　　　　Five.

(The teacher writes *5*.)

Ronnie took three of the cupcakes. How many did Ronnie take?　　　　Three.

(The teacher writes *5 – 3*.)

How many cupcakes are left?　　　　Two cupcakes are left.

(The teacher writes *5 – 3 = 2*.)

This is one way to write about the story we just acted out. It says, "Five take away three is two. Five cupcakes take away three cupcakes leaves two cupcakes."

When you are writing the equations to describe the stories, at various times use both the vertical and horizontal forms so the children become familiar with both. Begin reading the equation using the terms *plus*, *minus*, and *equals* once the children seem confident with the natural language. Explain that using these terms is just another way of reading the equation. Use the natural language (e.g., four and two is six or four jars of paint and two jars of paint made six jars of paint altogether) interchangeably with the formal language (e.g., four plus two equals six).

DISTINGUISHING BETWEEN THE PLUS AND MINUS SIGNS

The following games help children practice distinguishing between the plus and minus signs and performing the appropriate actions. The emphasis is on the signs and actions, and no totals are to be determined. Teach the games at first using the terms *get more* and *take away*, then introduce the words *plus* and *minus* as other ways of saying the same things.

Roll and Count

Materials: Plus and minus spinners—see p. 216 for directions for making • Die • Working space paper for each child—see p. 210 for directions for making • Unifix cubes

The children take turns rolling the die and turning the spinner. Each child adds or subtracts cubes from his or her working space paper according to the die and spinner.

For example:

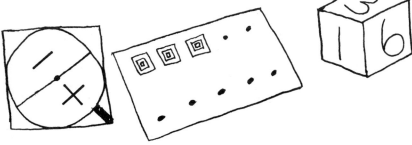

That's a plus, and I rolled a three—plus three.   Another plus.

I need two more—plus two.

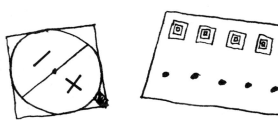

I got a minus. I rolled a three. That's minus three.

If the spinner and die indicate they are to take away more cubes than they have on their papers, the children say "impossible" and spin again. If they are to add more cubes than they have room for on their working space papers, they each get an additional paper.

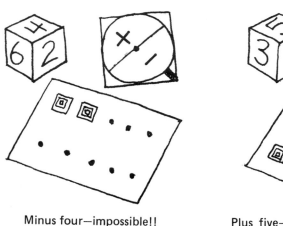

Minus four—impossible!!   Plus five—we need to get another paper.

Listen and Count*

Materials: Unifix cubes • Working space papers for each child • Plus and minus spinner (see p. 216) • Bell (or Xylophone)

The children take turns turning the spinner to determine the process to be performed. The teacher rings the bell to indicate the number of cubes to be added or taken away from the working space papers. If the number to be taken away is larger than the number of cubes on the papers, the children say, "Impossible!" If the number to be added is more than the space left on their working space papers, each child gets another paper.

Ding, ding.   Ding, ding, ding.

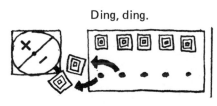

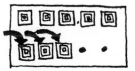

Minus two.   Plus three.

*Based on MATHEMATICS *THEIR* WAY, p. 190.

Ding, ding.

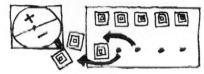

Minus two.

Grow and Shrink

Materials: Unifix cubes • Working space papers • Die

The children take turns rolling the die to determine what number they are to build on their working space papers. Each time they build a new number they verbalize what they have to do to make the new number while you write the appropriate symbols.

For example:

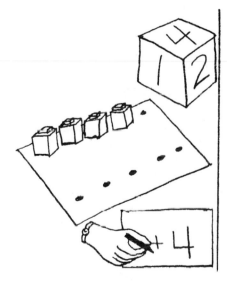

We rolled a four. We need to get four. Plus four.

(The teacher writes +4.)

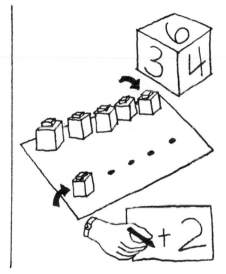

We rolled a six. We need two more to make six— That's plus two.

(The teacher writes + 2.)

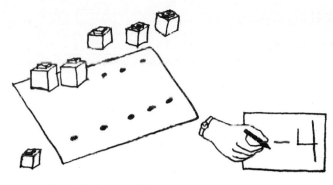

We rolled a two. We need to take four cubes off. That's minus four.

(The teacher writes – 4.)

Note: The die now indicates not how many to add or subtract as in "Roll and Count" but rather what number to make. Some children will have difficulty switching from "Roll and Count" to "Grow and Shrink;" however, it is essential that children keep flexible in their thinking and learn to deal with number in a variety of ways. It is important to play both these games, but be sensitive to the potential confusion and help the children by reminding them how the game is played as often as necessary.

ACTING OUT STORIES TO GO WITH EQUATIONS*

Materials: Unifix cubes sorted by colors • Counting boards or working space papers (non-dotted side)

After the children have had several experiences seeing you writing equations to describe stories, write an equation on the board and have them act it out with the cubes and tell their own stories.

*Based on MATHEMATICS THEIR WAY, p. 217.

For example, write *3 + 2 = 5*.

Nicole: I have three white kitties and two black kitties. That means I have five kitties.

Daniel: There were three bears in the woods and two more came. Now there's five bears in the woods.

Be sure to include subtraction equations as well so the children have frequent opportunities to distinguish between the two processes.

For example, write *6 – 2 = 4*.

Ryan: I saw six fish in the lake. I got two fish and then there were four left.

Christopher: There were six firefighters on the fire truck. Two firefighters went home. Four were still there.

WORKING INDEPENDENTLY WITH EQUATION CARDS AND COUNTING BOARDS

Materials: Unifix cubes sorted by colors • A set of counting boards for each child (8 identical boards)—see p. 210 • Addition and subtraction equation cards—see p. 214

After children have had experiences distinguishing between addition and subtraction signs in a directed group setting, they can begin to work independently with equation cards and counting boards.

A child chooses a set of counting boards, a container of cubes, and the appropriate equation cards. She or he spreads out the eight boards, puts an equation card with each board, and, after deciding what he or she wants the cubes to represent, puts them out on the boards according to the equations.

My cubes are fish. I got three fish, then two more came.

etc.

*Have the children put the cubes taken away next to the appropriate numeral on the equation card.

Note: Introduce the activity during a teacher-directed session so you can give help to those who need it. (You may need to spend a few days helping the children before they are able to work independently.)

Because the goal is for the children to be able to distinguish between addition and subtraction operations, they will be served best by having both processes to deal with from the beginning. However, since their attention initially will be on what to *do* and not so much on the concepts they are working with, it may be helpful to introduce the children first to one process and then the other.

Don't worry if there is some confusion; it is a natural part of learning. Children need a chance to sort things out, as long as there is help available if they need it. Even if you were to do one process (traditionally addition) followed by the other process (traditionally subtraction) for long periods of time, eventually you would still face the usual problem of children confusing the two processes. The children will be better served in the long run if they sort out these processes early, before getting locked into one or the other.

Once the children are able to work independently, the counting board activities can be monitored easily by a student tutor, a parent volunteer, or an instructional aide. Try to spend a moment now and then yourself just walking by to observe the children as they work. Resist the temptation to stop and teach; just note who needs help so you can provide the instruction needed at a more appropriate time. You will want to use this time to gather information on as many children as possible.

Children can work with the counting boards and equation cards over and over again. The boards provide in a more meaningful way the drill practice usually presented in the form of worksheets. The children maintain their interest because they can choose a variety of settings and pretend the cubes are many different things.

LEARNING TO WRITE EQUATIONS TO LABEL ADDITION AND SUBTRACTION STORIES

Materials: Unifix cubes • Counting boards

When you wish to have the children begin writing equations, continue to have them act out addition and subtraction stories as before. Provide them with individual chalkboards on which they can write the appropriate equations as the stories are being acted out.

Level One: Write the appropriate equation to describe the action in the story being acted out. The children should copy what you write.

Level Two: The children (without the help of a model) write the equation that describes the story being acted out. You then write the equation on the board so they can check the equations.

For example:

There were five apples on the tree.

Johnny picked two.

How many are left?

 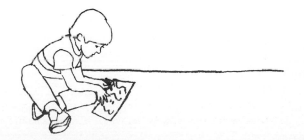

WRITING EQUATIONS INDEPENDENTLY

After the children have had practice writing equations in a teacher-directed situation, they can begin writing equations independently using the equation cards and counting boards. This activity can be done over and over again; in a sense, it replaces drill and practice with worksheets. The children will remain interested, because each time they work with the boards they can choose different settings and pretend the Unifix cubes are people, objects, or animals.* The activities can be done at the following three levels.

Making Equation Books Using Equation Cards

Materials: A set of counting boards for each child • Unifix cubes • Equation cards • 2 x 6 pieces of paper

The child chooses a set of counting boards and gets a container of Unifix cubes, equation cards, and 2 x 6 pieces of paper. She or he spreads out the counting boards and equation cards and places the Unifix cubes on the boards according to the cards. He or she then writes each problem and its answer on a piece of paper. When all the problems are completed, the papers are stapled together into a little book.

There are two apples on top of my tree and four more.

etc.

Making Up Equations

Materials: A set of counting boards for each child • Unifix cubes • 2 x 6 pieces of paper

The child no longer uses the equation cards but makes up his or her own problems. You can limit the number of cubes used by telling the child to use no more than ten cubes for each problem. If any children know how to write larger numerals and in your judgment do not need limits imposed, allow them to make up any problems they want.

Six children went playing in the grass. Two went home. Four are left.

Writing Stories

The children can write stories to go with the cubes and counting boards. If they are not ready to write independently, have an aide, parent, or student tutor help them.

The stories can be placed with the counting boards so that other children can read and act out the stories.

---

*Children using the counting boards and equation cards can be working side by side with children who are working with easier tasks such as counting out sets of cubes onto the counting boards or matching groups and numerals. Having children use the same materials at a variety of levels helps build an acceptance of individual differences.

For example:

(The child makes up a story and writes it.)

One child reads a story that has been written by another child, acts it out, and writes the equation.

Connecting Symbols to Number Combinations

After many experiences working with the combinations of four, five, and six as described in Section I, the children will be ready to learn to write the symbols for these combinations. If you introduce the writing of the combinations in a problem-solving setting, you can help children see that symbols are useful tools for keeping track of information. Have the children try to find all the possible combinations for the number being worked with. Record the combinations as the children report them.

Once the children are able to write the combinations themselves, they will be ready to work with the independent activities in this section on page 117.

When they are ready to work with writing the combinations of seven, eight, nine, and ten, they no longer need the teacher to model the writing so they should then work with the independent activities.

Teacher-Directed Activities

LEARNING TO RECORD THE COMBINATIONS

Present the familiar games described in Section I ("Number Combinations") in a problem-solving context, and record the equations as the children report them. Challenge them to look for all possible combinations as they work with the activities. Pose a question to the children.

Snap It

Today we are going to use six cubes to make a train. We want to find all the ways we can to break our train in two pieces. (It's okay to break zero cubes off.) Who knows a way we can break our train in two?

As the children report the ways they find, record the equation on the chalkboard.

For example: Nancy: I broke my train, so I have one and five.

This is how we write that

Continue to write the ways that the children report they have found. Each time they report a combination, they should look to see if it has already been written on the chalkboard.

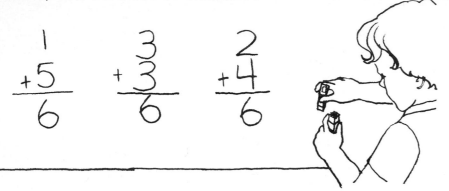

$$\begin{array}{r} 1 \\ +5 \\ \hline 6 \end{array} \qquad \begin{array}{r} 3 \\ +3 \\ \hline 6 \end{array} \qquad \begin{array}{r} 2 \\ +4 \\ \hline 6 \end{array}$$

I made three and three. Oh, I see it. It's already on the board.

On succeeding days, continue to explore the combinations using the other familiar activities. Model the recording process for the children.

The Tub Game

Today we are going to work with six cubes. We want to find out how many ways we can put some of our cubes on top of our tub and some of our cubes under our tub. Who can find a way?

$$\begin{array}{r} 3 \\ +3 \\ \hline 6 \end{array}$$

The Wall Game

How many ways can we make walls with our hands? Yes, five and one. We write:

$$\begin{array}{r} 5 \\ +1 \\ \hline 6 \end{array}$$

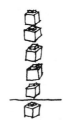

The Cave Game

How many different ways can we put cubes into our caves?

$$4 + 2 = 6$$

The Number Shapes

How many different ways can we put red and green cubes on our number shapes?

$$2 + 2 + 2 = 6$$

After many days of searching for the ways to make a number using all the various games, some children will be able to predict the ways that will be found. Some children will approach the task in an organized manner and eventually will find the pattern for all the possible ways. Allow the children to discover this in their own time. You can help them focus on what is happening if you ask questions such as:

How many ways did we find to make six? Do you think we've found all the ways?

How many ways did we find to make five?

Can you predict the number of ways we will find to make seven?

LEARNING TO WRITE NUMBER COMBINATIONS

When you wish to have the children write number combinations, provide them with individual chalkboards. Continue to have them search for various ways to make the combinations.

Level One: As you write the combinations that the children report to you, the children *copy* what you write.

Level Two: The children write the combinations as they are reported. After the children have finished writing, write the equation so they can check their work.

Have them record on the following worksheet the combinations found.

For example:

I found another way. Three on top and three underneath. That is three plus three.

Independent Activities

As soon as the children are able to write the number combinations with ease, they can work with the following activities.

HOW MANY WAYS?

Materials: (Will vary according to activity) Unifix cubes • Worksheets (see black-line master 50)

The following activities (previously explored with the teacher) can be experienced independently, using different numbers from those used earlier. The children will search for all possible combinations for the number being worked with.

For example: *How many ways can you arrange six cubes?*
using the tubs? using the "six" number shape?
playing the Cave Game? playing the Wall Game?

BUILD-A-FLOOR RACE

Materials. Build-a-Floor game board– run off black-line master 49 on ditto paper or tagboard • Unifix cubes • Dice • Worksheet—see black-line master 50

The children play the game as described on page 100. When they are finished, each one records his or her own combinations made on the following worksheet. (Do not require the use of the worksheet every time the game is played.)

Build-a-Floor game board

UNIFIX TRAIN OUTLINES

Materials: Unifix cubes (2 colors) • Crayons and pencils • Worksheets—see black-line masters 41–48, or 52

The children create Unifix train number combinations. They color the train outline to match the real Unifix train and write the appropriate equation.

The combinations can be created in a variety of ways:

a. The children arrange cubes of two colors in any way they wish and record the results. (Addition)
b. The children roll dice to indicate how many cubes to place on the outline. They fill in the rest of the outline with a second color cube and record the results. (Addition)
c. The children fill the train outline with cubes of one color. They roll the dice to determine the number of those cubes to take away and record the results. (Subtraction)
d. The children work in pairs. One partner makes up a problem using the cubes, and the other partner records the equation. (Addition and/or Subtraction)

Worksheets 41–48 or 52 can be used:

The child is given the whole worksheet; the assignment is to complete the worksheet.

or

The worksheet is cut apart into individual trains. The children are to do all they can and staple their papers into little books.

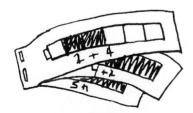

Or the children circle the train being used, then record the combinations on Worksheet 52.

NUMBER SHAPES

Materials: Unifix cubes • Crayons and pencils • Worksheets—see black-line masters 34–40, 51

The children create number combinations using the number shapes. They color the shapes on worksheets and write the appropriate equations.

The combinations can be created in a variety of ways:

a. The children arrange cubes of two colors in any way they wish and record the results. (Addition)
b. The children roll dice to indicate how many cubes of one color to place on the number shape. They fill in the rest of the shape with a second color cube and record the results. (Addition)
c. The children fill the number shape with cubes of one color. They roll the dice to determine the number of cubes they are to take away. The record the results as they go. (Subtraction)
d. The children work in pairs. One partner makes up a problem using the cubes, and the other partner records the equation.

Worksheets 34–50 or 51 can be used:

The child is given the whole worksheet; the assignment is to complete the worksheet.

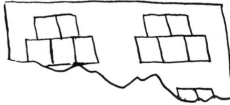

or

The worksheet is cut apart. The children are to do all they can and staple their papers into little books.

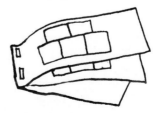

Or the children circle the number shape being worked with, then record the combinations on Worksheet 51.

WHAT'S MISSING?

Materials: Unifix cubes (one color) • Number shape • Individual chalkboards

Partner A makes up a problem for Partner B, using cubes and the number shape. Partner B records the equation.

Partner A fills the number shape with Unifix cubes.

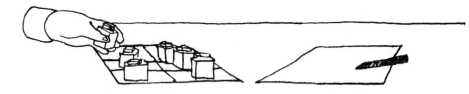

Partner B closes his or her eyes while Partner A removes some of the cubes.

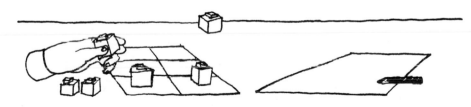

Partner B opens her or his eyes and writes the equation to tell what happened.

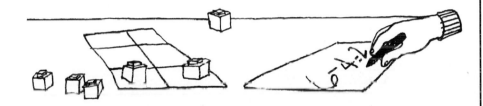

The partners then switch roles.

GRAB BAG WITH PARTNERS

Materials: Unifix cubes • Paper sack • Individual chalkboards or Worksheets—see black line master 50

The children play in pairs. Partner A fills the bag with the appropriate number of cubes, depending on what number is being worked with. Partner B reaches into the bag and takes out some cubes, showing Partner A what has been removed. They predict how many cubes they think are left. Then they check their predictions, and each child records the equation on a chalkboard or on worksheet 50.

Partner A   **Partner B**

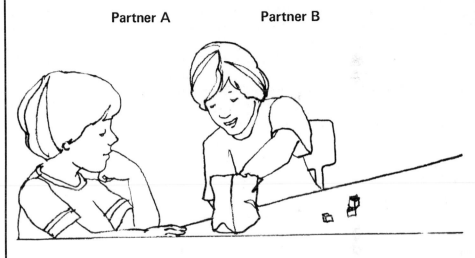

We had five cubes in the bag.

I think there are three left.
Let's check and see.

NUMBER TRAIN GRAPH

Materials: Unifix cubes (2 colors) • Butcher paper graph • Dittoed Unifix train outlines—see black-line masters 41–48 • Crayons

The following activity can be worked on by a group of children without direct teacher involvement. The children try to find all the possible arrangements for the combinations that make up a number being worked with. As they find the arrangements, they color a number train outline and place it in the appropriate column of the butcher paper graph. (The columns can be labeled with the possible combinations, or the children can label the columns as needed.)

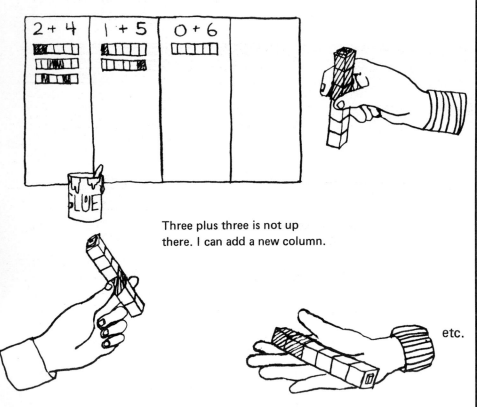

Three plus three is not up there. I can add a new column.

I have a different way to make one plus five.

Oh, one like this is already up. I can't use this one.

etc.

*Based on MATHEMATICS *THEIR* WAY, *Newsletter*.

REBUILDING WITH UNIFIX CUBES*

Materials: Unifix cubes (2 colors) • 18 x 24 newsprint folded into 8 spaces or 12 x 18 newsprint folded into 4 spaces

Build designs in each box for the assigned number. Write a number sentence to describe each design.

1 + 4 + 2	3 + 4	3 + 1 + 3	3 + 1 + 3

Remove the cubes from the paper. After reading the number combinations, rebuild—this time using the Unifix train format.

1 + 4 + 2	3 + 4	3 + 1 + 3	3 + 1 + 3
4 + 3	1 + 1 + 5	5 + 2	4 + 3

etc.

(On another day, the children can start with trains and rebuild with arrangements.)

Variation: Rebuild, using such easily available materials as toothpicks and bottle caps.

Writing Sums and Differences Independently

The following activities are designed to give children practice adding and subtracting in a variety of ways. They should be able to write equations independently before being assigned these tasks.

The children should be introduced to these tasks in a teacher-directed setting. Once introduced, each activity can be done over and over, because the worksheets are open ended and the children can create their own problems.

UNIFIX PUZZLES

Materials: Unifix cubes • Puzzles—see p. 211 for directions for making • 2 x 6 pieces of paper

The children fill in the puzzle with Unifix cubes and write equations to describe the puzzles. Some puzzles can be described in more than one way. Each different equation should be written on a separate piece of paper.

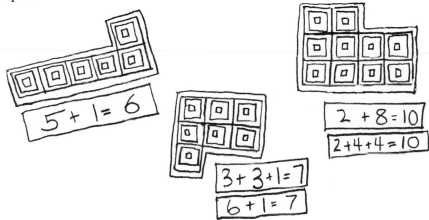

The papers can be stapled into booklets when they are complete.

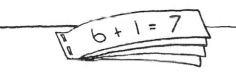

PARTNER PROBLEMS

Materials: Will vary   see below

Partners will take turns making up various problems for each other, changing the number of cubes used for each problem. If you want to confine the children to sums of six or less (or ten or less), tell them they can take only six (or ten) cubes. They can then use some or all of the cubes for each problem.

- One partner holds up a Unifix train of two-colors, and the other child writes the appropriate equation on a chalkboard or piece of paper.
- One partner holds up two Unifix trains (one in each hand), and the other child writes the appropriate equation.
- One partner puts some cubes on top of a margarine tub and some under a tub. The other partner writes the equation.
- One partner holds up a Unifix train, then breaks some cubes off. The other child writes the appropriate equation.
- Partner A puts some cubes into a paper sack, and partner B puts in some cubes. Both partners write the combination and predict how many altogether. The cubes are dumped and they count to find the total.
- Partner A puts some cubes into a grab bag. Partner B takes some out. The partners write the equation, predicting how many are left. The cubes are dumped and the partners check their predictions.
- Partner A puts cubes of one color on a working space paper. Partner B removes some of the cubes. They both write the equation.

COMBINATION DICE TOSS

Materials: Unifix cubes • Game board (teacher-made from 10 x 22 paper, marked off in 2" squares, labeled 0–10 at top of game board) • 2 dice (0–5)

Children roll two dice and build the number stacks to match. They place the two stacks in the column according to the sum. The game is over when one number wins (i.e., one column is filled). This game provides a visual display of the various combinations that make up a particular sum.

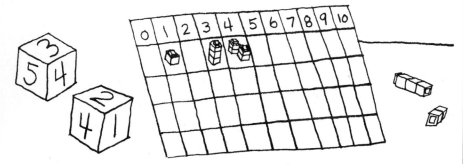

Three and two make five.
I put these by the five.

Extension: The children can record the combinations on a worksheet (see black-line master 50).

For example:

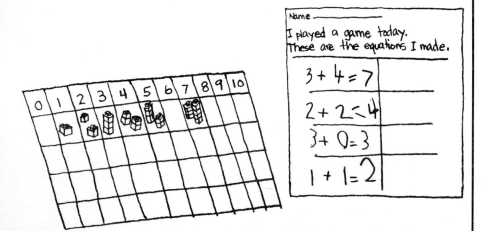

Name _____
I played a game today.
These are the equations I made.

$$3 + 4 = 7$$
$$2 + 2 < 4$$
$$3 + 0 = 3$$
$$1 + 1 = 2$$

ADDITION AND SUBTRACTION SPIN IT

Materials: Various number shapes • Unifix cubes • Plus and minus spinner--see p. 216 for directions for making • Paper

Have available a variety of number shapes with which the children have already worked. A child chooses one of the number shapes and turns the spinner to see if he or she is to add or subtract. The child performs the appropriate action and records the equation.

For example: The child picks a number shape.

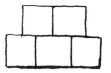

She or he turns the spinner, which lands on *plus*.

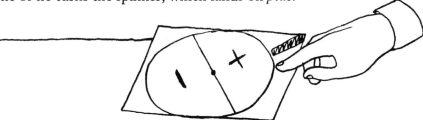

The child puts cubes of two colors on the number shape and writes the equation.

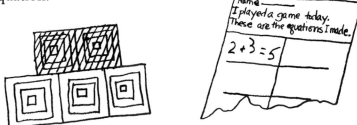

Name _____
I played a game today.
These are the equations I made.

$$2 + 3 = 5$$

He or she chooses another shape and turns the spinner again. The spinner lands on *minus*.

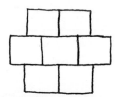

The child fills the number shape with cubes of one color and then takes some cubes off. She or he writes the equation.

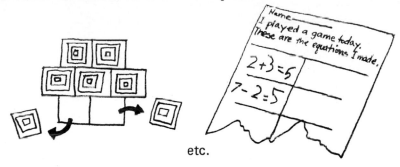

etc.

Variation: Use number train outlines instead of number shapes.

MAGIC BOX EQUATIONS

Materials: Magic box and Magic box cards—see p. 213 for directions for making • Unifix number line —see p. 217 for directions for making • Unifix cubes • Paper

When a card is placed in the top slot of the magic box, it will turn over so that the back of the card appears when the card comes out of the bottom slot. Children can make up equations using magic box cards.

The children choose a magic box card and place the number of cubes indicated by the card on the Unifix number line. They write the numeral on their papers.

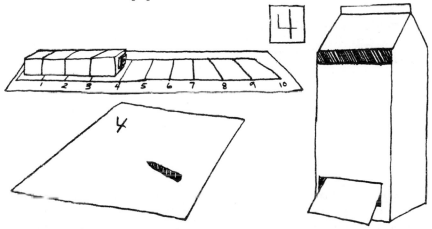

*Based on MATHEMATICS *THEIR* WAY, p. 246.

The children place the card into the magic box and add or subtract cubes to the number line to make the number that comes out of the box. If the number is larger, they add and complete the written equation.

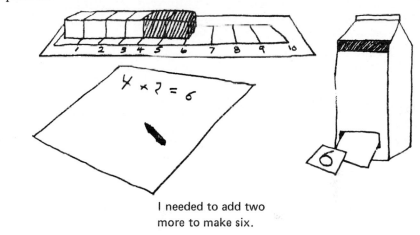

I needed to add two
more to make six.

If the number is smaller, they subtract and complete the written equation.

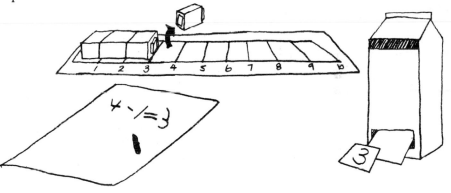

I needed to take one
off to make three.

WHAT NUMBERS CAN YOU MAKE?*

Materials: Unifix cubes • Crayons • Worksheets (see black-line masters 53, 54)

The children make five Unifix stacks (each stack made with different color cubes and no stack more than five cubes high). There can be more than one stack of any particular number. The object is to try to make each of the numbers from one to ten in as many ways as possible. The children are allowed to use more than one stack to make any number, but they may not break any stacks apart. If they can't make a number, they are to cross it out. If they can make the number, they color in the outline on the worksheets and write the equation that describes the trains used.

For example:

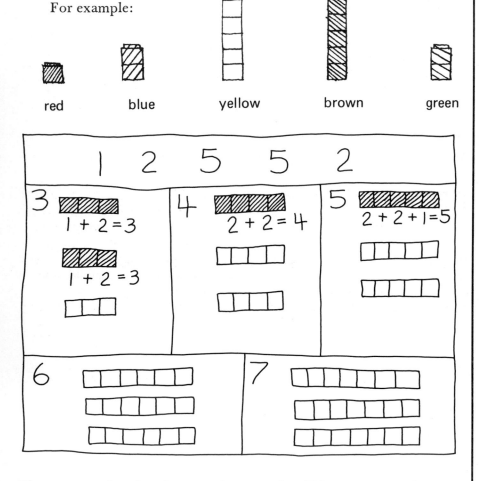

The game can be played over and over, as it will be a new experience each time the numbers used changes.

Variation: The children start with five stacks, as described previously. However, this time they record the combinations they can make on a graph (see black-line master 1).

For example:

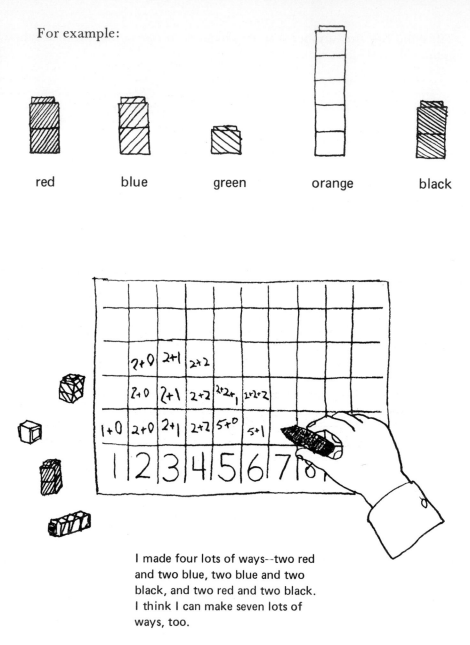

I made four lots of ways--two red and two blue, two blue and two black, and two red and two black. I think I can make seven lots of ways, too.

WORKBOOK PAGES/DRILL SHEETS

Materials: Selected workbook pages • Unifix cubes • (Other materials will vary according to the activity chosen)

Once children understand the processes of addition and subtraction and the way symbols are used to record these processes, the workbook pages can be used to provide opportunities to practice writing answers to addition and subtraction problems. If you have the children use the familiar activities to determine the answers in the workbook, you will help them make the connection between the concepts they are learning and the format of the workbook.

For example:

Margarine tubs

Unifix trains

Working space papers

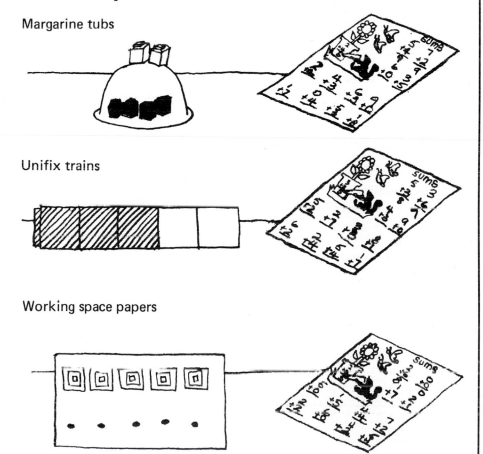

Counting boards

Note: Have the children choose *one* way to complete a page.

ANALYZING AND ASSESSING YOUR CHILDREN'S NEEDS

The assessments presented in this book are different from many presented elsewhere, because this method is concerned with not only whether children can write answers to addition and subtraction problems but also the level of understanding they have of the concepts represented by the symbols. What children write down is only a small indication of what they are thinking. You can learn most about your children by observing them while they are working. If you observe children carefully and note how they respond in a variety of settings, and if you try to understand why they respond the way they do, you will become more and more proficient at providing the right experiences for them.

You can observe several children working on the same task and see that each is experiencing that task at many different levels. For example, Randy, Leslie, and Craig are arranging six Unifix cubes into as many designs as they can. Randy has made the following arrangement. The teacher tries to point out the number combination that is in his arrangement.

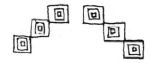

Look, Randy. There is three and three.
No, that's six. See. One, two, three, four, five, six.

Randy's focus is on counting out the right number of objects. He is not yet ready to see or deal with the smaller numbers that combine to make the larger numbers. As Randy has experiences with real numbers and sees and hears them described by their parts, he will discover the smaller numbers contained in the larger ones and learn that numbers can be described by a variety of names such as three and three, four and four, or six and zero. This will happen naturally as long as the opportunity is there for him to experience this concept. At this point, it would be inappropriate to urge him to memorize his basic facts.

◆ ◆ ◆

Leslie has made several arrangements.

The teacher stopped to talk to her about her work.

Oh, I see you made a design with three and two and one.
Yes, and I made a two and two and two. And here is three and three, and this is three and three and one. And I can make more, too.

Leslie is very intrigued with all the ways that six cubes can be arranged. She is very comfortable describing them by their parts, and she is focused on the fact that numbers can be described in lots of ways but is not trying to remember combinations. She as yet has not noticed the inconsistency in thinking three and three is six and three and three and one is six; she just assumes she has discovered another way. The teacher can help Leslie notice her counting error by having her count several of her designs and allowing her to discover her mistake. Given enough experiences with combinations of six, she will begin to recognize the same combinations coming up again and again, and, given enough time, she will internalize the combinations of six.

◆ ◆ ◆

Craig also has made several arrangements.

The teacher asked him if he could describe his arrangements with numbers.

Yes. There's two and three and one, five and one, four and two, six and zero. *Can you think of any number combinations you haven't made yet?*

Craig thought a moment and said:

Oh yeah, I didn't make three and three yet.

Craig knows the combinations for six. This activity is appropriate for Craig because of the problem-solving nature of the activity. He is challenged by the task of finding all the possible ways. He knows the combinations for six already, however, so he does not need to play the more structured games like "Snap It" and "The Cave," where the combinations for six are confronted over and over again. Craig needs to move on to larger numbers for which he does not know the combinations. The hiding assessment described later in this section will help in determining the number(s) he still needs to work on.

◆ ◆ ◆

A great deal of information about your children is available to you without having to do formal assessments. The problem for you as a teacher is to know what to look for and how to interpret what you see.

This section is designed to help you become more aware of the kinds of behavior you can expect from your children and to aide you in determining the needs indicated by their behavior. To help with this, suggested assessment procedures have been included. These have been presented to help you tune in to the level of your children's thinking about addition and subtraction, their sense of number quantities and relationships, and the way they deal with symbols. The procedures are not designed as a list of objectives that can be mastered or not mastered.

You can use the assessment questions in a variety of ways:

- As a guide to help you know what to look for as you watch your children work.

- To assess some of your children individually in a formal setting and from this experience learn what to look for in the less formal situations.
- To assess those children for whom you need more information than you have been able to get from watching them while they work.
- To formally assess all your children in order to have written information to present to parents as you explain your emphasis on concept development rather than on memorization of symbols.
- To assess children at the beginning of the year, or when you have a new student and you want to know where to begin.

HIDING ASSESSMENT

This assessment will provide you with the information you need to assign the appropriate number or range of numbers to your children when working with number combinations (see Section I: Number Combinations, pages 83-108).

Ask a child to hand you five cubes. Hide some of them in one hand and show the child the other cubes. Ask, "How many are hiding?"

Change the number of cubes hidden and ask again, "Now how many are hiding?" Repeat for several combinations.

Do not show the number of cubes that had been hidden nor indicate by your facial expression whether the child is right or wrong just take note of the responses.

If the child is successful with five, check larger numbers in the same way. If the child is unsuccessful with five, check smaller numbers.

Children's Responses

When Karla's teacher was assessing her knowledge of the combinations of five, Karla would make totally inappropriate responses. She would guess that eight or ten were hiding, even though the total num-

ber of cubes being used was five. When she was assessed at three, however, she knew the combinations quickly and confidently. When assessed at four, she again gave inappropriate responses.

The numbers beyond four or five do not have much meaning for Karla. Because her responses were so far off for four and five, the teacher suspects Karla will need lots of time to work with the combinations for four and five (reviewing three occasionally to help her see how the numbers relate to each other). The teacher will probably want to introduce Karla to the number combination games, using the number *three* so Karla can focus on how the activity is played and not with learning the combinations at the same time that she is learning the games.

◆ ◆ ◆

Paul could figure out the combinations for five and six by using his fingers but was not confident. When assessed at four, he knew the answers instantly.

Because Paul could figure out the answers to the combinations for five and six—even though he did not know them instantly—he will probably need less time than Karla to become really confident with the range of numbers from four to six and will be ready to deal with the symbols for these combinations before too long.

◆ ◆ ◆

Jamie knew the combinations for five instantly. The teacher assessed him for six and seven, and he was equally successful. When working with eight cubes, he had to stop and figure out the answers using his fingers to help him, and he sometimes came up with the wrong answer.

Jamie is ready to work with the combinations for eight, nine, and ten. His teacher already knows that Jamie has not yet learned to write addition and subtraction equations, so the teacher plans to teach him the symbols for the combinations he already knows. When he is comfortable writing the equations for four, five, and six, he can begin exploring the larger numbers.

NUMBER TRAIN COMBINATIONS

The following assessment is designed to give you several kinds of information about your children in a short amount of time. Rather than in-depth information, you will end up with some clues to the way a child deals with the following number concepts.

- Instant recognition
- Counting on
- Comparing numbers
- Combining numbers
- Writing equations

Show a child a Unifix train made of two colors and arranged in the following manner.

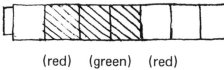

(red)   (green)   (red)

How many red? How many green? Which color is more? Which color is less? How many more _____ than _____ ? How many altogether? Write a number sentence that tells about how many are in the train.

If the child is successful, repeat the questions, using a longer train arranged in the following manner.

Children's Responses

red and green train

When asked to tell how many red and then how many green, Peter counted by ones to find out. He knew there were more red cubes than green cubes, but answered that there were four more red, telling the total number of red cubes. When asked how many altogether, he counted all the cubes and arrived at the right number. When asked to write a number sentence, he said he didn't know how.

Peter is working with number at a basic counting level. He does not recognize groups of three or four instantly and needs to count to find out that four and three make seven. He does not understand the question "How many more?" Peter's teacher should provide him with lots of activities from all the sections in this chapter. He should begin focusing on numbers to six and gradually move to working with numbers above six.

When shown the red and green train, Joni said without counting that there were four red and three green cubes. When asked how she knew so fast, she said you can tell just by looking. She knew there was one more of the red cubes than the green cubes. When asked how many altogether, she said, "There's four and three more. I think that's seven." When asked if she could write with numbers to tell about the Unifix train, she wrote a 7.

When the teacher showed Joni the longer yellow and blue train, Joni said there were four blue cubes. She counted, "Three, four, five, six" to determine the number of yellow cubes. When asked how many more yellow than blue cubes, she put out six fingers and said there were three more yellow. When asked how many in the train altogether, she said, "Six yellow and four blue makes . . . six, seven, eight, nine, ten. Ten altogether."

Joni already has a good sense of number. She recognizes small groups of cubes instantly and knows that three and four are seven without counting to find out. When she doesn't know an amount, she counts on to find out. She knows what the question "How many more?" means and can figure out the answer if she doesn't know it.

Joni knew that three plus four equals seven, but her teacher still wants to assess her (using the hiding assessment) to see if she needs to work with numbers larger than seven to learn those combinations. She did not know how to write a number equation to describe the Unifix train. The teacher will want to check further to see how Joni deals with symbols in other settings. She may know how to read symbols but not write them, or she may not be at all familiar with addition and subtraction equations. If the teacher finds that Joni does not know how to work with symbols, she will have Joni begin working with the activities that connect symbols to the concepts (Section II).

yellow and blue train

WORKING WITH SYMBOLS: WORD PROBLEMS

The following assessment is designed to find out if children can listen to a word problem and write a number equation to describe what happened. The assessment will also determine if they can make up a story when given an equation.

I am going to tell you a story, and I want you to write a number sentence that tells about the story. You can use the cubes if they will help you.

There were three cars in the parking lot. Two more cars came. How many cars are there altogether?

Six children were playing in the sandbox. Two children left. How many children are still playing in the sandbox?

Here is a number sentence. Make up a story to go with it.

$$3 + 4 = 7$$

Here is another number sentence. Can you make up another story?

$$4 - 3 = 1$$

Children's Responses

When presented with a word problem, Misty says, "I don't know how." The teacher realizes that Misty needs experiences described in Section I, p. 79, even though she has shown that she can do some addition and subtraction problems.

When asked to write an equation to go with a word problem, Dennis wrote:

$$3 + 2 - 5 \qquad \begin{array}{r} 6 \\ - \\ 2 \\ = \\ 4 \end{array}$$

When asked to make up a story to go with an equation, he told a simple but accurate story for both addition and subtraction equations. Dennis could figure out what number to write but needs to practice the correct format (see Section II, p. 113).

Frances wrote both equations correctly and was able to tell a story for both addition and subtraction equations. The teacher does not need to provide Frances intense instruction in word problems at this level. She needs to give Frances opportunities to work with word problems on occasion throughout the year. As Frances begins to work with place value and multiplication and division concepts, the teacher will provide word problems where larger numbers and other processes are used.

WORKING WITH SYMBOLS

Give the children a paper on which the following equations are written:

5 + 4	8 – 3
2 + 3	6 + 3
5 – 1	3 + 6
7 + 1	4 + 2
9 – 1	6 – 4

Point to the *5 + 4* problem and say:

Show me with the cubes.
Write your answer.
Read it to me.

Point to the *8 – 3* problem, and repeat the above questions.

Then tell the children to do the rest of the page. Ask them if they can do any of these without the cubes and without their fingers. If necessary, they can use the cubes.

Children's Responses

When the teacher pointed to the *5 + 4* problem and asked Ginny to show this with the cubes, Ginny counted out five cubes. Then she counted out four cubes. The teacher asked, "What do you have?" Ginny said, "I have five in this pile and four in this pile."

Ginny's teacher then pointed at the *8 – 3* problem and said, "Do this one." Ginny put out eight cubes and then got three more. Ginny did not even seem to notice the plus and minus signs. She saw the numerals and counted out a group of objects to match. Just to make sure, the teacher pointed to each of the signs and asked Ginny if she knew what they were. Ginny said she didn't know. She was not asked to do the rest of the page.

Ginny can count accurately and is ready to begin working on addition and subtraction activities from this chapter.

◆　◆　◆

Don put out five cubes and then four cubes. He pushed them together and counted to nine. He then wrote *9*.

5 + 4 9 (He left out the equals sign.)

He then put out eight, got three more cubes, counted them all, and wrote *11*.

8 – 3 11

He read, "Five makes four is nine. Eight makes three is eleven."

Don's teacher asked him to look at the other problems on this page. She said, "Can you do any of these without counting the cubes and without counting your fingers?" Don looked at the paper and shook his head.

The teacher pointed to the *2 + 3* and *5 – 1* problems and said, "Do just these two problems. You can use the cubes." Don put out two and then three cubes, counted them all, and wrote *5*. He then put out five more cubes, got one more cube, and wrote *6*.

The teacher then asked Don, "If you had five oranges and I gave you one more, how many would you have?" Don said, "I don't know." His teacher asked him to figure it out. He put out five fingers and one more and counted to six. The teacher then said, "If you had two cars and your brother gave you two more, how many cars would you have?" "I don't know. I have to count," said Don.

Don does not have a well-developed sense of number. He is confused about the plus and minus signs and adds for either. Through lots of experiences with the activities from all the sections in this chapter, he will develop a stronger sense of number and will learn how the symbols record his number experiences.

◆　◆　◆

When Norma was shown the equation *5 + 4,* she said, "That's nine." The teacher said, "Show me." Norma put out five and then four cubes, counted and said, "See. Nine." The teacher said, "Show me this one" and pointed to the *8 – 3* problem. Norma said, "That's five." The teacher again said, "Show me." So Norma put out eight cubes. She then got three more cubes and took those three cubes away. She looked puzzled when she saw that there were eight left instead of the five she had said, so she tried again. The next time she

placed the three cubes to the right of the eight cubes and then took them away. When she again ended up with eight cubes, she said, "I guess eight take away three is eight." She wrote *8* as the answer to the equation.

8 – 3 = 8

Norma has learned to add and subtract with symbols, but when she is asked to demonstrate what the symbols mean, she can show addition but does not really understand subtraction. Even though Norma has memorized answers to subtraction problems, she still needs to experience subtraction using the Unifix cubes. She needs to learn to label the experiences with the appropriate symbols until the symbols are connected in her mind to what they represent.

◆　◆　◆

Mike showed 5 + 4 correctly with the cubes and wrote = *9*. He then did 8 – 3 correctly and wrote = *5*. He wrote the answers to 2 + 3, 5 – 1, 7 + 1, and 9 – 1 without hesitation. When he had to do 6 + 3, he said, "I don't know this one" and he got out the cubes. He looked at the next problem, 3 + 6, and asked, "Is that the same as six plus three?" The teacher said, "What do you think?" Mike said, "I think so, but I'm not sure." He put out three and then six cubes, counted them, and said, "I thought so." He wrote the answer to 4 + 2 but was not sure of 6 – 4, so he got out the cubes to find out.

Mike has a good sense of number and understands the processes of addition and subtraction. He is beginning to see the relationship between 6 + 3 and 3 + 6 but does not yet see that 4 + 2 and 6 – 4 are related.

Mike's teacher will want to assess him with the hiding assessment to see what number combinations he needs to work with. He will want to give Mike more experiences with numbers to ten, but he will want to choose those activities that will encourage Mike to think and that will help him discover number relationships.

◆　◆　◆

It is not always practical for teachers to spend long periods of time assessing individual children in formal settings. Again, you will learn more watching your children work than performing formal assessments. Many of the assessments just described can be done when you ask a whole group to do an activity by focusing in on how one or two children deal with the problem.

It is also important to remember that concept development happens gradually rather than in nice, sequential steps. Try as you might, you cannot know with great precision or certainty exactly what each child is thinking and understanding. Children's levels of development and thinking is in constant flux as they have new experiences.

Your goal is not to produce automatons that spit out answers on cue but rather children who are confident with the world of number, who have faith that numbers make sense, and who feel they as individuals can make sense of that world of numbers. In the end, they devise their own systems for dealing with the tasks you present. You are obligated not to know exactly what the next exact step should be but rather to provide the kinds of experiences that allow children to create the meaning in their own ways.

CHAPTER FIVE

If your textbook or workbook objectives
are:
- Naming digits in the ones,
  tens, and hundreds places
- Writing to 100 (and beyond)
- Comparing two-digit numbers
  to determine which is more
  and which is less
- Skip counting
- Addition of two-digit numbers
  (with or without regrouping)
- Subtraction of two-digit numbers
  (with or without regrouping)

Then you are dealing with:

Place Value

WHAT YOU NEED TO KNOW
ABOUT PLACE VALUE CONCEPTS

The concepts related to the understanding of place value are not easy for young children to grasp. Many children exposed to these ideas too soon, too fast, and/or too abstractly remain uncertain and confused throughout their elementary school experience. Year after year, children ask their teachers, "Do we have to borrow?" "Is this when we carry?" Children diligently memorize steps and rules for getting answers only to forget them or misapply them. The teacher's job in primary grades is to help provide the kinds of experiences that build a foundation for understanding so that children will be able to make sense of the rules they learn and determine for themselves that they are following proper procedures for getting reasonable answers.

In order to provide the kinds of experiences that will lead to understanding, you should be aware of the following notions that children must deal with in coming to terms with place value concepts.

FORMING AND COUNTING GROUPS

The most basic concept children must confront is that our number system is based on the formation of groups of ten. When we have ten units we group them into one group of ten; when we have ten groups of ten, we regroup them into one group of one hundred; when we have ten groups of one hundred, we regroup them into one group of one thousand, etc.

Counting groups requires a different kind of thinking from counting single objects. Children's first counting experiences require an understanding of one-to-one correspondence. They learn that one number word goes with one object. But when dealing with numbers above ten, they are required to count groups as though they were individual objects. The question "How many tens in thirty-four?" assumes the child can conceive of ten objects as one entity. The question "How many hundreds in 346?" assumes the child can conceive of one hundred objects as one group (all the while remembering that each hundred is also ten groups of ten).

CHANGING VALUE OF NUMERALS

Another key idea the children must learn is that a particular digit can stand for many different amounts, depending on its place in a number. Imagine for a moment what it would be like if every number we could think of had to be written with a unique symbol: the thought is overwhelming.

We can all be grateful that a system was devised that requires us to learn only ten different digits, which can be put together in ways that allow us to write any number—no matter how large. This system works because the value of each of the numerals changes according to the size group it stands for. The size of the group is indicated by the numeral's position in the number.

As wonderful as this system is, it is quite a step for children who have just recently learned that 7 stands for a particular number of objects to be asked to understand that 7 also stands for varying amounts, depending on its position. For children who still don't see the difference between 7 and ⟨ or *saw* and *was*, who are still learning to tell their left hands from their right hands, this can be quite a mystery.

This concept is made even more complicated when one number can be represented correctly in several ways but is incorrect if written in slightly different ways. *Seventy* can be represented as 7 tens and 0 ones, 7 tens, or *70* but not as *7* or *07*. Seventy-eight can be represented as *78, 70 + 8,* 7 tens and 8 ones—even 8 ones and 7 tens—but not as *87* or *708*. To many children, these differences are very subtle.

PATTERNS IN THE NUMBER SYSTEM

Once we have discovered the patterns in the number system, the task of writing to one hundred (and beyond) is simplified enormously. What seems potentially so confusing and difficult is much easier to deal with because the number system is so orderly and predictable. We encounter the same sequence of numbers (0, 1, 2, 3, 4, 5, 6, 7, 8, 9) over and over again. Children do not see the patterns automatically so will need experiences that allow them to discover these patterns.

CONSERVATION OF LARGE NUMBERS

In Chapter One we discussed the idea of conservation of number (see page 1). Children have conservation of number when they realize that the quantity of a group of objects does not change, even when those objects are moved closer together, farther apart, or hidden or rearranged in some way. Although children may have developed conservation for small numbers, it is not unusual for children as old as eight not to have developed conservation for large numbers. It is not obvious to these children that twenty-six cubes grouped into two tens with six left over is still twenty-six. For some it appears to be less (or more). For others the action of rearranging or regrouping seems like an operation such as addition or subtraction, and they think they have ended up after regrouping with a number different from the one they started with.

ADDITION AND SUBTRACTION

Addition and subtraction of numbers above ten are different from addition and subtraction of numbers less than ten, because children must deal with the notion of forming and counting groups and the changing value of numerals while they are joining and separating quantities. The process and thinking that is required is quite complex.

THE USE OF MODELS TO
TEACH PLACE VALUE CONCEPTS

The use of manipulative materials such as Unifix cubes can help children in dealing with place value concepts. The unique qualities of Unifix cubes can make them especially effective tools.

The children can physically join ten cubes into a single unit so that ten single objects become one object. Joining these ten cubes into one train does not cause the units to disappear. The children can still count and check, if necessary, to see how many units they have. This activity is a beautiful model of the idea that the quantity of ten can be one ten and ten ones at the same time.

If a child starts with fourteen loose cubes and snaps together ten of these into a train, all that has changed is the *arrangement* of the cubes. The child still has the same fourteen cubes he or she started with. Unifix cubes potentially can help children develop conservation of number in a way that is not possible with an abacus or blocks of various sizes designed to represent units, tens, and hundreds.

Some math materials used to teach place value concepts require children to trade ten units for one object (such as a larger block, a colored chip, etc.). When children are just beginning to figure out what regrouping is all about, trading can be an extra step that clouds their understanding of the concept. The original units they started with are gone, and there is no way for them to verify that they still have what they started with; they simply have to take a teacher's word for it.

There is a time when trading for one object that represents many objects is appropriate. It can be very convenient when dealing with very large numbers; however, such materials should not be used too soon.

Section I: Developing
the Concept of Regrouping

Three main strands are presented in this section.

THE GROUPING GAMES: The activities presented here are designed to help children understand the process of grouping and regrouping. They begin the exploration of this idea by working with small groups of four, five, and six. When working with these small groups, the need to regroup comes up quickly and often, thus giving the children lots of practice working with the regrouping notion. Exploring this idea with each of the numbers four, five, and six makes it easier for children to distill the idea of regrouping as a process that is not dependent on the size of the group.

DISCOVERING NUMBER PATTERNS: Next, the children will discover the number patterns that occur when they record each step of the process of forming groups. When working with four, five, and six, the number sequences are shorter and easier for the children to see than the long number sequences that occur when working with groups of tens.

WORKING WITH GROUPS OF TEN: All of the introductory activities provide a foundation for understanding base ten. Once children understand the concept of regrouping and learn to find the patterns that evolve, they then explore these same ideas working with tens.

Many of these lessons in Section I are highly structured, with the entire class following each step of the teacher's instructions. Because of the nature of those lessons, it seems most helpful to present them in script form.

It is important that you include all your students in the introduction of the place value grouping games, even though you may feel that not all of them are ready for place value activities at the same time or that not all should move through the place value activities at the same pace. All the children should be in on the choosing of the nonsense words (described in the following pages) so that when they are ready for more experiences with these activities, they will know where the words come from.

The Grouping Games*

The children explore the idea of regrouping by working with groups of four, five, and six. The children as a class pick a nonsense word to name each group. Naming gives a group an identity and helps the children think of it as one entity. Referring to a group of objects by name is less confusing than using a number word like *four* or *five*.

INTRODUCING THE GROUPING GAME

Materials: Unifix cubes (sorted by colors) • Place value boards (1 per student)—see p. 212 for directions for making

Today we are going to play a counting game. In this game, we can't say the word four. *We have to make up a new word that means* four. *The word can't mean anything else. Who has an idea?*

Zib. (This is just an example of a nonsense word. Your students will invent their own word.)

Zib. Yes, that's a good name to use for our game. I will write it on the chart so we won't forget.

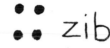 zib

Once a nonsense word is picked, it will be the class word for a group of four whenever the place value counting games are played with groups of four.

When we play the zib game, this is how we will count. One, two, three, zib. Help me count these books, and we'll make some zibs.

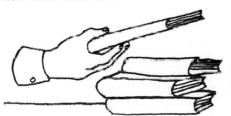

One, two, three, zib.

One, two, three, zib.

Let's count ourselves. Every time we get a zib, the children in the zib will hold hands.

one   two   three   Zib!   etc.

Help me count these Unifix cubes.

one ▢

two ▢ ▢

three ▢ ▢ ▢

zib ▭▭▭▭

When I say "zib," that means I have to snap the cubes together and make one zib.

How many zibs am I holding?

One zib.

*Based on MATHEMATICS *THEIR* WAY, p. 276.

Pass out place value boards to each student, and make Unifix cubes available. (It is helpful if the cubes have been sorted by color, this eliminates children digging for favorite colors or trying to make patterns when you are trying to keep the activity moving.)

This is called a place value board. Place it so that the happy face is smiling at you.

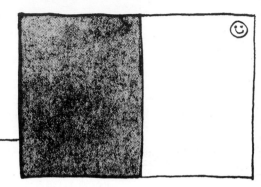

Put your hand on the white side. Now find the blue side. The blue side is for the zibs and the white side is for the loose cubes. How many zibs do we have on our board now?   Zero.

How many loose ones?   Zero.

Yes, we have zero zibs and zero. When I say "plus one," I want you to put one cube on the white side of your board. "Plus one." How many do we have now?

Zero zibs and one.

(Initially, you will be saying "zero zibs and one, etc." with the children until they pick up on what they are supposed to say and are able to say it without you.)

Plus one. How many?

Zero zibs and two.

Plus one. How many?

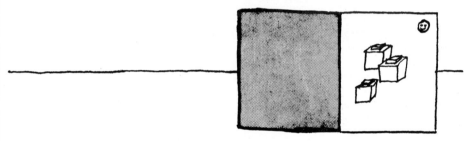

Zero zibs and three.

Plus one. How many?

Zero zibs and oops, that's a zib.
We have a zib.

What do you think we have to do when we have a zib?

Yes. Where do we put zibs?

Now how many do we have?

Do we have to snap it together?

On the blue side of the board.

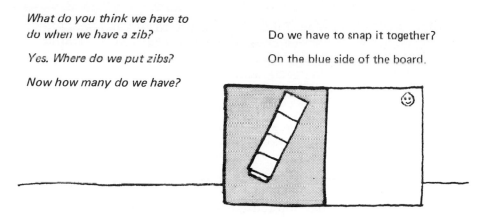

One zib and zero.

Continue saying "plus one" until the children have three zibs and three on their boards.

In the beginning, it will be easier if the children do not have to regroup beyond two places. If you were to continue to say "plus one" after the children have three zibs and three on their boards, the children would have another zib. If they then put that zib on the zib side of the board, they would have four zibs. Whenever the children reach four zibs, they must regroup the four zibs into another group, placing the zibs into a container (we call this new group a big zib). Big zibs belong in the space to the left of the board.

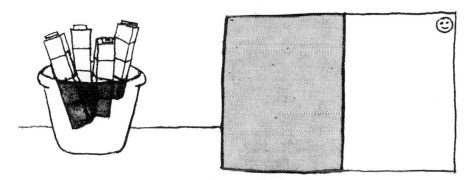

To avoid dealing with big zibs in these first lessons, stop saying "plus one" when three zibs and three is reached. See pages 142–143 for a description of what to do when you are ready to extend the regrouping beyond two places. [1]

Instead, when three zibs and three is reached, tell the children that you are going to start saying "minus one." Ask them what they think they will have to do. Many will figure out that they will need to remove one cube each time.

Now I want you to begin taking cubes off your board when I say "minus one."

Minus one. How many?

Three zibs and two.

Minus one. How many?

Three zibs and one.

Minus one. How many?

Three zibs and zero.

Minus one.

We don't have any more.

Yes, we do. We have zibs. We can take one off the zibs.

(When you reach this point, talk it through with the children so they see the logic in what they have to do instead of only learning a procedure.)

If you take one cube off a zib, is it still a zib?

No. There's only three.

Only zibs belong on the blue side— so what shall we do?

Put the three cubes on the white side.

Yeah, but first we have to break them up.

How many zibs do we have now?

Two zibs and three.

Minus one.

Two zibs and two.

(Continue saying "minus one" until you reach zero zibs and zero.)

You do not want your children to think that the grouping game is to be played only with zibs. After two to five more lessons with zibs, begin playing the grouping game with other size groups as described on the following pages.

PLAYING THE GROUPING GAME WITH OTHER SIZE GROUPS

Materials: Unifix cubes (sorted by color) • Place value boards (1 per student)

In order for children to develop the concept of regrouping, they will need to experience this regrouping process with groups of various sizes. Therefore, after two to five lessons using groups of four, introduce them to grouping by fives. Have the children choose a nonsense word to stand for groups of five, and then add it to your classroom chart.

For example:

We are going to play a new counting game. In this game we can't say "five." We need a new word. Who has an idea?

Glom.

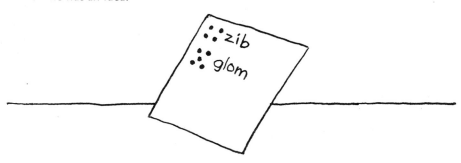

(Remember, the word *zib* was a label for groups of four and is not used when counting gloms.)

Play the glom game with the place value boards and Unifix cubes, going to four gloms and four and then back to zero gloms and zero.

After a few days, repeat the procedure for groups of six. Choose a new word for groups of six, and add it to the class chart. Using the place value boards and Unifix cubes, play the plus-one game until reaching five groups and five, and then play the minus-one game until you reach zero groups and zero.

After the children are comfortable with each of the groups you have introduced, spend a few days working with any of the groups in random order until you feel the children are able to regroup easily, no matter what size group is being used. Be sure you play both the plus-one and minus-one games.

Although it is not necessary to explore groups of numbers other than four, five, and six in order for children to understand regrouping, many children become intrigued with the process and wish to explore these larger groups. Allow those children to work with groups of seven, eight, nine, or whatever size they want.

Extension: Rather than adding or subtracting one cube each time, have the children add and subtract various numbers of cubes.

For example:

We are working with zibs today. Listen carefully, because I'm going to say different numbers. And sometimes I'll say "plus" and sometimes I'll say "minus."

Plus two. How many?

Zero zibs and two.

Plus three.

We have enough for a zib.
But there's one left over.

How many?

One zib and one.

Plus two. How many?

One zib and three.

Plus three.

We have another zib.

How many?

Two zibs and two.

Minus three. How many?

One zib and three.

REGROUPING BEYOND TWO PLACES

Materials: Unifix cubes • Place value boards (1 per student) • Margarine tubs

During the initial regrouping experiences, it is enough for young children to learn to form groups, take groups apart, and place the cubes in the correct places, using the two spaces on the place value board. It is important, however, that children do *not* get the idea that the grouping ends with just those two places. They need to see that regrouping occurs every time the number being worked with is reached, no matter what size group is being worked with.

You can help children get this idea by working with small groups of three, four, or five. Because the numbers are small, the need to regroup occurs often, and the pattern for the process becomes apparent quickly. Later, when the children work with tens, hundreds, and thousands, the process already will be clear to them.

For example, when working with groups of four (zibs, in our examples), and you have four cubes on the white side of the board, they must be regrouped into one zib.

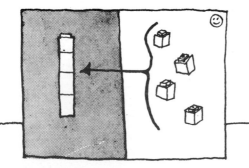

When you have four zibs, they must be regrouped into one big zib.

Add a piece of paper for each new area.

When you have four big zibs, they must be regrouped into one super zib.

When you have four super zibs, they must be regrouped into one gigantic zib.

etc.

Discovering Number Patterns

Once the children can make groups easily, you can help them see the number patterns that occur.

DISCOVERING THE NUMBER PATTERNS FOR PLUS-ONE GAMES*

Materials: Unifix cubes (sorted by color) • Place value boards • Paper for recording patterns (see below)

Play the counting games in the usual way, but this time write each step of the pattern as the children report the number of cubes they have on their boards at each step.

We are going to play the glom game today. When you tell me the number of cubes you have on your board, I am going to write the numbers down.

(Write the numbers on a long strip of paper that has been shaded on the left side with a blue crayon. A roll of paper towels works well, or use shelf paper or butcher paper.)

How many cubes are on your boards right now?

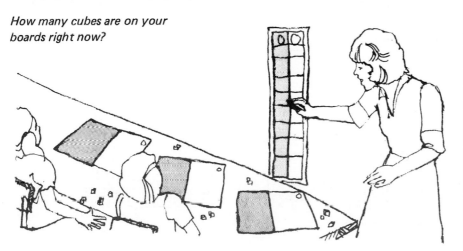

Zero gloms and zero.   Write:

Plus one. How many?

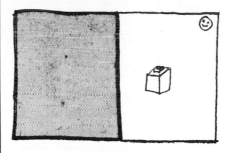

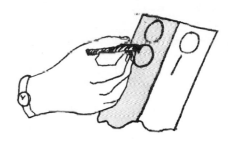

Write:

Zero gloms and one.

Plus one. How many?

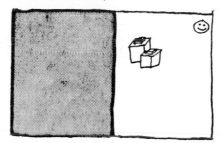

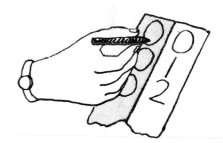

Write:

Zero gloms and two.

Continue until you reach four gloms and four. When you and the children have reached four gloms and four, have the children look at the numbers on the white side of the strip and say them with you. The patterns will be easier for the children to discover if you tell them to clap each time they say "zero."

Zero, one, two, three, four; zero, one, two, three, four; etc.
(clap)   (clap)

Loop the number patterns as they are repeated, as shown on the next page.

*Based on MATHEMATICS *THEIR* WAY, p. 299.

Next, have the children look at the shaded side of the strip. Have them say the numbers and stand or sit as the numbers change. Loop the patterns as the children say them.

Zero, zero, zero, zero, zero; one, one, one, one, one; two, two, etc.
(stand)   (sit)   (stand)

After the patterns have been looped, have the children count the number of zeros, the number of ones, etc., on the shaded side of the strip.

Children: There are five zeros.   etc.

Extension: On succeeding days, repeat this activity, using the other groupings, and compare the various number patterns that occur. Rather than recording the patterns for the children, you can have them work independently, recording these patterns on strips and looping the patterns (see black-line master 68).

DISCOVERING THE NUMBER PATTERNS FOR THE MINUS-ONE GAMES

Materials: Unifix cubes • Place value boards • Paper for recording patterns

Allow children the opportunity to discover the reverse pattern that occurs when you remove cubes from the board. Have the children put the appropriate number of cubes on their boards (in the case of zibs, it would be three zibs and three). Record the number of cubes on the boards at each step, and look for the patterns that occur.

For example:

Put three zibs and three on your board.

Write:

Minus one. How many?

Write:

Minus one. How many?

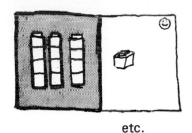

etc.

Write:

When you reach zero zibs and zero, have the children read the pattern while you make the loops. When the children are looking for the pattern on the white side of the paper, have them clap each time the first step of the pattern is repeated. When looking at the shaded side, have them stand or sit as the pattern changes.

Extension: Have them play the minus-one game with the other groupings as well. They can work independently, recording the patterns on the strips and looping the patterns they find.

Extension: The Plus-Two, Plus-Three, Plus-Four, Etc. Games. Some children will want to see what patterns emerge when they add more than one cube at each step. Allow those students to explore these patterns. Many will want to go beyond the two places on the place value board. Show them how to make the "big groups" by putting cubes in a tub and how to record these groups by writing in the margins. (If working with other bases is a new experience for you, explore these patterns with your children. It's fun!)

For example:

I am working with zibs. I am adding two.

I am working with gloms. I'm adding three.

DISCOVERING THE NUMBER PATTERNS IN A MATRIX

Materials: Uniflx cubes · Place value board · Matrices (cut from 1″ graph paper) Use a 4 x 4 matrix when playing with groups of four, a 5 x 5 matrix when playing with groups of five, a 6 x 6 matrix when playing with groups of six, etc.

Play the counting games with the children, and use a matrix to record the number patterns that occur.

For example:

Today we are going to play the zib game. As we play, I want you to tell me what to write.

How many on our board now?

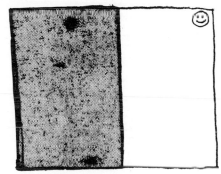

Zero zibs and zero.

Teacher writes:

Plus one. How many?

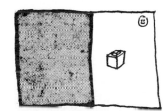

Zero zibs and one.

Teacher writes:

Plus one. How many?

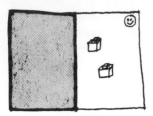

Zero zibs and two.

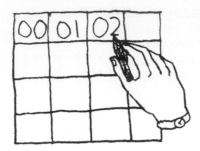

Teacher writes:

(Continue until the matrix is completed.)

After the matrix has been filled in, have the children look for the patterns that occur in the matrix.

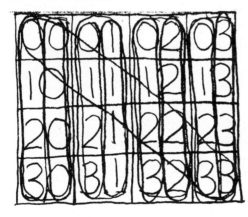

Extension: On succeeding days, use the other groupings. Give the children the appropriate matrix, and have them work independently, doing their own recording.

Working with Groups of Ten

The previous activities have built a foundation for understanding. The children have experienced the idea of regrouping and have seen the patterns that emerge. Working with tens now is just another way to do what they have already done.

INTRODUCING GROUPING BY TENS

Materials: Unifix cubes • Place value board (see p. 212)

Introduce the children to the idea of ten by saying:

Today we are going to work with a new group. The group is ten. Instead of making up a special word for ten, we are just going to call the group with ten in it a ten. (Add the word *ten* to your chart.) *We are going to play the plus-one game. How many do we need to have before we can put any on the blue side?*

Ten?

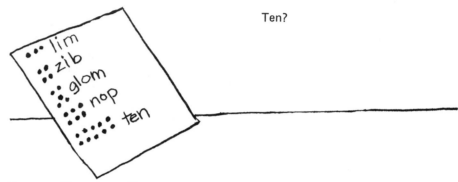

Yes, we will need ten. How many on your board now?

Zero tens and zero.

Plus one. How many?

Zero tens and one.

Plus one. How many?

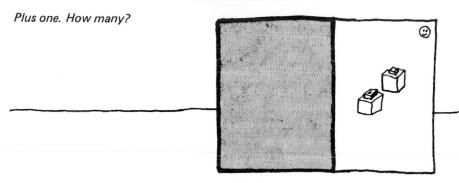

Zero tens and two.

(Continue until you reach about three tens and four.)

By this time it will be obvious to you why working with the smaller numbers was so useful in introducing the important concepts. It will take quite a while for you to reach even three tens and four with the children, and they will have practiced grouping only three times!

WRITING THE BASE TEN PATTERNS
INDEPENDENTLY—Writing the Ten Patterns on a Strip

Materials: Place value strips—see black-line master 68 · Unifix cubes · Clear tape
3″ x 4″ tagboard for each child, to be taped into a roll.

Have the children play the plus-one game independently, and have them write the numbers on the place value strip.

When a child finishes a strip, have him or her loop the patterns on the strip and then tape another strip to the completed one.

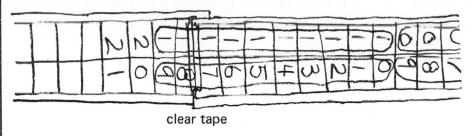

clear tape

The child then can tape the strip to a cardboard or tagboard roll, wrap the strips around the roll, and attach a wooden clothespin that has her or his name written on it.

The children should continue playing the plus-one game, writing the numbers, and adding to their strip until they teach *at least* 150. This will take more than one day. Many children get very involved in this task and like to see how far they can go. Encourage them to go as far as the materials allow. At the end of a period, the children should roll up their strips and put the cubes that have been snapped into tens into a container without breaking them apart. The next day, they should read the last number on their strips, get the necessary tens and ones to build the number on the place value board, and continue working.

Exploring the Base Ten Patterns in a Matrix

Materials: Worksheet—see black-line master 69 · Unifix cubes

Have the children play the plus-one game in base ten, and record the numbers in a 10 x 10 matrix.

The result will be a 00–99 chart that the children will have created themselves. Have them look for patterns in this chart.

Note: Looking for patterns in the 00–99 chart can be a fascinating activity that can be repeated many times. Once the children have made their own chart, hang a large one up in the classroom that is laminated or covered with acetate. On occasion, discuss the patterns with the children, and loop the patterns on the chart.

Ask such questions as:

> What number am I covering up? How do you know?
> Do you see a pattern in the columns? in the rows?
> What patterns do you see in the diagonal lines?

MARGIE'S GRID PICTURES

Materials: Unifix cubes • Grid picture task cards (teacher-made, see examples) • 00–99 chart (black-line master 70) or 10-by-10 grid with first column of numerals filled in—see black-line master 69

Children can become familiar with the 00–99 chart by working with grid picture task cards. They create pictures by placing Unifix cubes directly on a 00–99 chart or on a 10-by-10 grid which has only the first column of numerals filled in.

For example:

```
                task card 1

red:    4, 13, 14, 15, 22, 23, 24, 25
        26, 31, 32, 33, 34, 35
        36, 37, 40, 41, 42, 43
        44, 45, 46, 47, 48, 50
        52, 54, 56, 58

black:  64, 74, 84, 94
        82, 92, 93
```

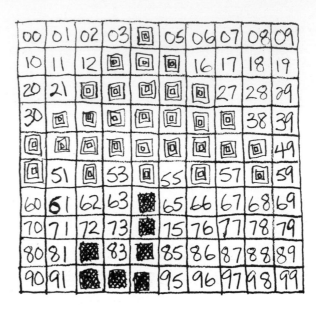

It's an umbrella!

Examples of grid picture task cards:

(tree)

green   4, 5, 13, 14, 15, 16, 23, 24, 25, 26
        32, 33, 34, 35, 36, 37
        42, 43, 44, 45, 46, 47, 51, 52
        53, 54, 55, 56, 57, 58, 61, 62
        63, 64, 65, 66, 67, 68

brown   74, 75, 84, 85, 94, 95

(horse)

black 11, 12, 20, 21, 22, 30, 31, 32
43, 44, 45, 46, 47, 48, 53, 54
55, 56, 57, 59, 63, 64, 65, 66
67, 72, 73, 77, 78, 81, 82, 83
89, 90, 91

(turtle)

green 35, 36, 44, 45, 46, 47, 53, 54
55, 56, 57, 58, 40, 41, 50, 51
62, 63, 64, 65, 66, 67, 68, 69
72, 79

(flower)

orange 11, 12, 21, 22, 16, 17, 26, 27
51, 52, 61, 62, 56, 57, 66, 67

blue 33, 34, 35, 43, 44, 45

green 54, 64, 74, 84, 94, 85, 76, 93
82

(flag)

yellow 0
brown 10, 20, 30, 40, 50, 60, 70, 80, 90
blue 11, 12, 21, 22

red 13, 14, 15, 16, 17, 18, 19
31, 32, 33, 34, 35, 36, 37
38, 39, 51, 52, 53, 54, 55
56, 57, 58, 59

white 23, 24, 25, 26, 27, 28, 29
41, 42, 43, 44, 45, 46, 47, 48, 49

(ship)

blue. 80, 81, 82, 83, 84, 85, 86, 87, 88, 89
90, 91, 92, 93, 94, 95, 96, 97, 98, 99

brown 40, 41, 42, 47, 48, 49, 51, 52, 53
54, 55, 56, 57, 58, 62, 63, 64, 65
66, 67, 73, 74, 75, 76

red 4, 14, 24, 34, 44

orange 5, 6, 7, 8, 15, 16, 17, 18

(butterfly)

red 10, 20, 30, 40, 50, 60, 70, 80
21, 31, 41, 51, 61, 71, 32, 42, 52, 62
43, 53, 18, 28, 38, 48, 58, 68, 78, 88
27, 37, 47, 57, 67, 77
36, 46, 56, 66, 45, 55

blue 2, 13, 15, 6, 24, 34, 44, 54, 64
74, 84, 94

THE PLUS-TWO, PLUS-THREE, PLUS-FOUR, ETC., GAMES IN BASE TEN—Recording the Patterns on Strips

Materials: Place value strips (see black-line master 68)

Have the children add by twos, threes, fives, tens, etc., and record the patterns that evolve.

Recording the Patterns on a 00-99 Chart

Materials: Unifix cubes • Worksheets (see black-line master 70) • Crayons

The children add by twos, threes, fours, etc., and record the numbers they get by coloring in the appropriate number on the 00-99 chart.

For example:

	01	02	03		05	06	07		09
10	11		13	14	15		17	18	19
	21	22	23		25	26	27		
30	31		33	34	35		37	38	39

I am adding by fours.

	01	02	03	04		06	07	08	09
	11	12	13	14		16	17	18	19
	21	22	23	24		26	27	28	29
30	31	32	33	34	35	36	37	38	39

I am adding by fives.

Grab and Add

Materials: Unifix cubes • Place value strips • Dice (optional)

Have the children take (grab) a handful of cubes, arrange them into tens and ones, and place them in the appropriate spaces on the place value board. They then choose a number (or roll a die) and repeatedly add that number. They record the resulting number patterns. They will discover that patterns appear no matter what number they start with.

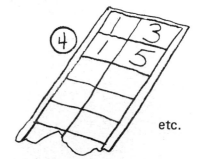

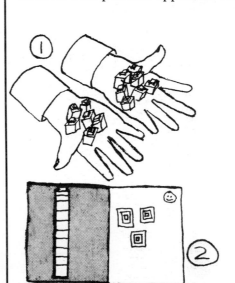

I had one ten and three ones.

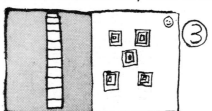

I will add two each time.

etc.

Section II: Developing a Sense of Quantities Above Ten

The following activities are designed to broaden children's understanding of large numbers and prepare them for addition and subtraction with regrouping. If enough time is spent helping children develop a strong sense of large numbers, you can avoid many of the problems that typically occur when children work with place-value ideas. Also, you will learn much about your children's development of these concepts as you observe them working with these activities.

Teacher-Directed Activities

The amount of time children will need to work with the following activities will vary widely. Watch your children closely to see how they deal with the questions you pose. Some will need fairly brief periods of time working with you before they understand the concepts. Others will need continually to review these ideas, even after they have been introduced to addition and subtraction.

ARRANGEMENTS

The following activities help the children discover that the number of cubes remains the same no matter how they are arranged. This concept, referred to as *conservation of large numbers,* is a key to understanding addition and subtraction with regrouping. This idea develops with experience over time so will not be learned by all children in a few lessons. Give the children a variety of directions that require them to arrange and rearrange cubes.

Arranging Piles of Loose Cubes into Tens and Ones

Materials: Unifix cubes • Place value boards—see p. 212 for directions for making

Make a pile of 26 cubes. Check your friend's pile of cubes while he or she checks yours so you can make sure you each have twenty-six cubes. Snap together ten of your cubes. How many tens do you have?

One ten.

How many loose ones?	Sixteen.

Raise your hand when you can tell me how many cubes you have altogether.

As surprising as it may seem, many children will not know that they still have the original twenty-six cubes. By having the children raise their hands rather than shouting out their answers, you can determine the various levels of thinking. You will see some children who are confident there are twenty-six cubes and raise their hands instantly, some who will start counting on from ten, and others who will need to count each of the cubes one by one.

Do we have enough loose cubes to make another ten?	Yes.
	I don't know.
Let's try to make another ten. Now how many tens do you have?	Two tens.
How many loose cubes?	Six.

How many altogether?

(As before, allow each child to determine the answer to that question in any way that makes sense to her or him, including counting all twenty-six cubes, one by one.)

Breaking Trains into Tens and Ones

Materials: Unifix cubes • Place value boards

Make a train that is thirty-two cubes long. Line it up with your neighbor's train so that you can make sure you both have the same number of cubes. How many tens do you think you will be able to make?

Three.

Four.

Two.

Break your train into tens and ones and see. How many?

Three tens, and we have two cubes left.

How many cubes do you have altogether?

We still have thirty-two cubes.

(Watch to see how the children determine this.)

Repeat, using other numbers.

Arranging Cubes in a Variety of Ways

Materials: Unifix cubes (some already grouped into tens) • Place value boards

Get thirty six cubes.

(Make note of who counts out thirty-six cubes and who gets three tens and six ones.)

How can you arrange these cubes on your place value boards?

We could put three tens and six ones on our boards.

Let's try that.

Did that idea work? How many cubes do you have altogether?

We still have thirty six.

Who has a different idea?

We could just have one ten and the rest loose.

Let's try that. How many loose ones?

Sixteen are loose.

Is there any other way we could arrange them?

They could all be loose.

Repeat, using different numbers.

Note: At first it may seem that having the children arrange cubes in a variety of ways is working against the idea you want them to develop. However, you want them to realize the same number of things can occur in a variety of ways. Later, they will find ways to arrange the cubes most efficiently (see "Build It Fast" on page 152).

Extension: Write symbols to describe the arrangements as the children build them.

For example:

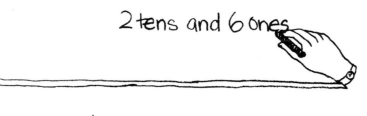

$$20 + 6$$

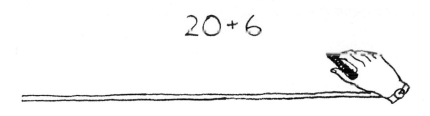

REARRANGE IT

It seems obvious that fifteen is one ten and five ones, that twenty-five is one ten and fifteen ones and that one ten and thirteen is twenty-three. But this is not obvious to children. The following activities will help them discover these relationships.

Focusing on Numbers Between Ten and Twenty

Materials: Unifix cubes • Place value boards

Make a pile of fourteen cubes. Who can predict how many tens we can make? How many ones do you think we will have left? Let's try it and see.

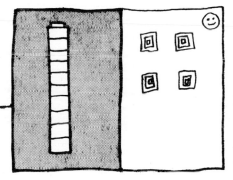

Repeat, using other numbers.

Note: Many children still will not be *sure* that fourteen will make one ten and leave four ones. It will take many experiences before children internalize this concept.

Focusing on Numbers More than Twenty

Materials: Unifix cubes • Place value boards

Put one ten and fifteen ones on your board. How many cubes do you think you have altogether? Check and see.

Clear your board.

Put two tens and thirteen ones on your board. How many cubes do you think you have altogether?

Note: Some children may still be counting by ones to find out. Allow those who found out quickly to share what they did with the others.

Breaking Tens into Ones

Materials: Unifix cubes • Place value boards

(This is a basic step in subtraction with regrouping. Children need to understand that the number of total cubes has not changed even though they have broken cubes apart.)

Put three tens and four ones on your board. How many cubes is that?   Thirty-four.

Break up one ten and put it with the ones. How many tens do you have now? How many ones? How many altogether?

Repeat, using a variety of numbers.

BUILD IT FAST

Materials: Unifix cubes (some already grouped into tens)

In the previous games, the children learned to be flexible in arranging cubes in a variety of ways (developing conservation of number). This game encourages them to arrange cubes in the most efficient manner (developing an appreciation of the usefulness of grouping by tens).

Present a variety of numbers to the children both orally and in written form. Have them build the numbers as fast as they can. Review periodically until children are applying this concept in other settings.

GIVE AND TAKE WITH TENS AND ONES

Materials: Unifix cubes • 12 x 18 pieces of paper

Place some cubes on a place value board so the children can see how many there are.

Cover the cubes with a 12 x 18 piece of paper. Reach under the paper to add or remove cubes. Have the children tell how many they think there are each time you add or remove cubes.

For example:

I'm putting one more ten on the board. How many do you think there are now?

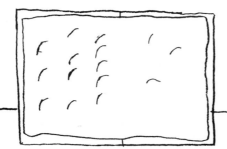

(Lift the paper to check.)

I'm taking one cube off. How many do you think there are now?

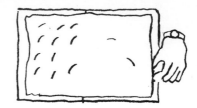

(Lift the paper to check.)

Extension: Have the children *write* how many at each step.

CAN YOU GIVE ME?

Before introducing children to the formal subtraction process, play the following game to help them distinguish between subtraction problems that are impossible and those that are possible but that require regrouping.

Put four tens and five ones on your board. How many is that?

Forty-five.

Can you give me sixty-two? (Show *62* written on a chalkboard while you say it.)

No. We don't have enough cubes.

Can you give me this many? (Show *34* written.)

Yes. We can give you three tens and four ones.

Okay. Put those back. How many cubes do you have?

Forty-five.

Can you give me forty-nine? (Show the written symbol.)

No, we don't have enough.

Can you give me twenty-nine?

No, we don't have enough ones.

Yes we can. Forty-five is bigger than twenty-nine.

How can you give me twenty-nine if you don't have enough ones?

We can break up a ten, and then we'd have more ones.

What do you have when you break up a ten?	We have three tens and fifteen ones. Now we can give you two tens and nine.
Okay. Put those back. How many cubes do you have now?	Forty-five.
Can you make me fifty-six?	Yes, we can break up a ten and get more ones.
	No, we can't. There still won't be enough.
Let's try it and see.	Nope. Impossible! It doesn't work. We don't have enough tens to take away five.

Repeat this game for a few days until the children are able to tell easily which numbers are possible and which are not.

INTERPRETING SYMBOLS

To give children practice relating the numerical symbols and the number of cubes represented by the symbols, write numerals on the chalkboard and have the children build the numbers.

For example:

Build this number on your place value board.

24

Don't be too surprised or dismayed if a few children put two cubes on their boards rather than two tens. It takes a long time and many experiences for some children to sort this all out. Help them focus on *what* the numerals stand for by having them read the numerals in tens and ones, calling twenty-four "two tens and four" and calling twelve "one ten and two."

Extension: Present numbers that require the children to think carefully about what the symbols mean.

For example: 40

04

4 tens

4 ones

3 tens and 4 ones

1 ten and 0 ones

0 tens and 1 one

40 + 5

30 + 8

Have them build the number with Unifix cubes to show they understand.

Independent Activities

Children need many experiences focusing on tens and ones while working with large numbers. The following activities provide such opportunities.

When you introduce the following activities, allow the children to determine in their own ways the number of cubes they used. After they have experienced some of the difficulties of large numbers, talk to them about ways they could organize their cubes so they can determine quickly how many tens and ones they have without having to count each cube one at a time. Tell them you also want to be able to walk by and tell how many cubes they have used without having to count each cube.

Some of the ways they will find to organize the cubes are to:

Change colors every time they get to ten.

a.

b.

Lay one down when they get to ten.

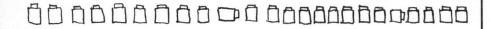

Mark each time they get to ten with one color cube.

Snap together ten cubes. Then snap together ten more cubes, not hooking the two trains together.

Sometimes, stop the children while they are working on the following activities and have them report to you the number of tens and ones they have used so far. They will begin to appreciate the organization that prevents them from having to count by ones all the time.

LOTS OF LINES

Materials: Unifix cubes • "Lots-of-Lines" task cards—make by drawing various lines on 9" x 12" pieces of tagboard; label each with a letter—see examples below • Cards labeled "Lots-of-Lines" so the children can write it on their worksheets • Worksheets—see black-line master 71

Have the children guess and then determine how many Unifix cubes will fit along the lines and record their results on worksheet 71. (The cubes can be loose or snapped together, depending on the type of line.)

For example:

(Remind the children to organize their cubes in some way. See p. 155.)

PAPER SHAPES

Materials: Various paper shapes cut from tagboard and each labeled with a letter (teacher-made—see examples below) • Unifix cubes • Worksheets—see black-line master 71 • Cards labeled "Paper Shapes" so the children can write it on their worksheets

Cut out various shapes from tagboard. Label each with a letter. The children estimate the number of cubes it will take to fill in a paper shape and record that number on their worksheet. Then they fill in the shape and write how many cubes they used.

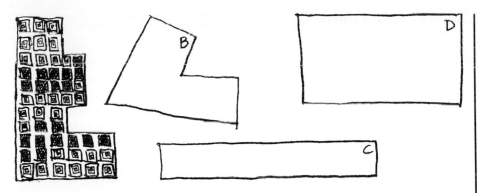

(Remind the children to organize their cubes in some way. See p. 153 .)

Variation: The children can cut their own paper shapes and label them with their initials.

CONTAINERS

Materials: Various containers (margarine tubs, tin cans, shoes, small boxes) labeled with letters • Unifix cubes • Worksheets—see black-line master **71** • Cards labeled "Containers"

Provide various containers. Have the children guess and then determine how many Unifix cubes the containers hold and record their results on worksheet 71.

For example:

Extension: Pose questions to some of the children:

*Does the container always hold the
same number of cubes? What can
you do to make it hold the most?*

YARN

Materials: Various lengths of yarn (labeled with letters written on masking tape) • Unifix cubes (sorted by color) • Cards labeled "Yarn" • Worksheets—see black-line master **71**

Have the children guess and then determine how many cubes long the yarn is and record their results on worksheet 71.

For example:

YARN SHAPES

Materials: Lengths of yarn labeled "A, B, C," etc. • Unifix cubes (sorted by color) • Worksheets—see black-line master 71

The children take a piece of the yarn used in the previous yarn activity and make a closed shape with it. They will estimate the number of cubes it will take to fill in the shape and record that number on their worksheets. They then fill in the shape with cubes and find out how many.

They take the same piece of yarn and make a different shape with it, estimate the number of cubes it will take to fill this new shape, fill it in with cubes, and write the number of cubes they use. They should use the same piece of yarn at least three different ways.

For example:

I'll make this shape with my yarn. I think it will hold 200 cubes.

I used 140 cubes.

MEASURING THINGS IN THE ROOM*

Materials: "Measuring Things" cards (teacher-made—see examples below) • Unifix cubes • Card labeled "Measuring Things" • Worksheets—see black-line master 71

The child chooses a Measuring Things card and then after guessing, uses Unifix cubes to measure the object indicated on the card. The guess and the results can be recorded on worksheet 71.

Extension: The children can compare things in the room and fill in the blanks in the following sentences.

*Based on MATHEMATICS *THEIR* WAY, p. 307.

(Sample sentence to be copied)

Note: If you do not want the children to work with numbers larger than one hundred, be sure you do not include objects in the measuring cards that are longer than one hundred cubes.

Ideas for Measuring Things Cards:

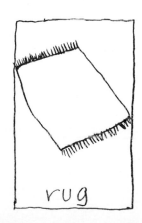

MEASURING MYSELF

Materials: Measuring-Myself Cards (teacher-made)—see example below • Unifix cubes • Worksheets—see black-line master 71 • Card labeled "Measuring Myself"

The child chooses a Measuring-Myself Card and then, using Unifix cubes, measures the body part indicated on the card. (Because some body parts are difficult to measure by oneself, children may wish to work with partners.) The results are to be recorded on worksheet 71.

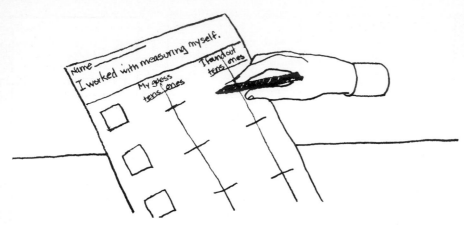

Extension: The children can compare body parts and fill in the blanks in the following sentences.

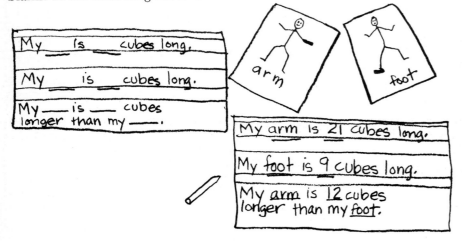

Ideas for Measuring-Myself Cards:

Front of card	Back of card	Front of card	Back of card

etc.

COVER IT UP

Materials: Unifix cubes • Worksheets—see black-line master 71 • Card labeled "Cover It Up" • Various things in the room • A list of these things

Have the children determine the number of Unifix cubes it takes to cover a variety of surfaces.

For example:

- A book
- My desktop
- A chair seat
- An outline of my foot
- The bottom of the wastebasket

Make a list of the possibilities, and add to it as children get other ideas of things to measure. The children can then use the list to help them spell and record their experiences on worksheet 71.

For example:

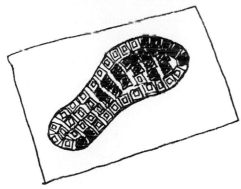

COMPARING MYSELF

Materials: Things-in-the-Room Cards • Unifix cubes • Worksheets—see black-line master 72

Have the children measure their heights using Unifix cubes. They can then measure various things in the room to determine if they are longer or shorter than themselves. They can list the things measured in the appropriate places on their worksheets.

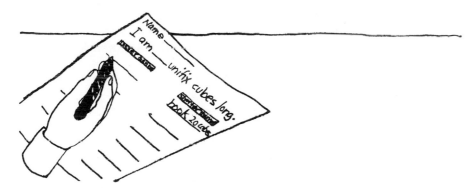

MAKING TRAILS

Materials: Unifix cubes (corner cubes optional—see p. 217) • Making-a-Trail Cards (teacher-made) see example below

The children choose a Making-a-Trail Card and determine the number of cubes it takes to make a trail from one place in the room to another, as indicated on the card. (Corner cubes may encourage exploration of different types of trails.)

For example:

The children then record the number of cubes it took by copying the sample sentence and filling in the blanks.

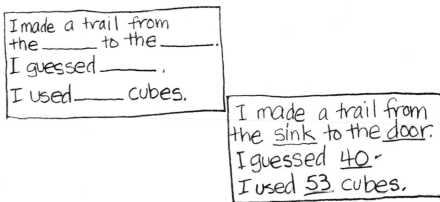

Pose a challenge for some by asking the following questions:

Do you always get the same number when you make a trail? What is the shortest trail you can make? the longest?

Variation: The children can make trails around objects such as a chair, a desk, or a wastebasket. Ask questions such as:

Is it farther two times around the wastebasket or once around my desk?

Note: The focus of this activity is not on accurate measurement of distance; therefore, the term *trail* is used rather than *distance from*.

Suggestions for making Make-a-Trail Cards:

From the sink to the door.

From the round table to the shelf.

From the orange rug to the wastebasket.

From the waterfountain to the chalkboard.

From the easel to the teacher's cupboard.

From my cubby to my friends cubby.

UNIFIX STACKS*

Materials: Unifix cubes • Build-a-City game board (run off on ditto paper or tagboard—see black-line master 33 • More/less spinner (see p. 217) • Dice (marked 4–9) • Worksheets—see black-line master 31

The partners take turns rolling the dice and placing the appropriate number of cubes on each section of their game board. When all sections of the board have been filled, players count their cubes by arranging them into tens and ones. Each person records both numbers on her or his worksheet.

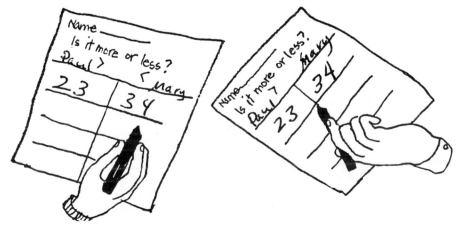

*Based on MATHEMATICS *THEIR* WAY, p. 320.

One partner turns the spinner to see if the person with more or the person with less cubes wins that round of the game. They circle the winning number and play again.

RACE TO ONE HUNDRED

Materials: Place value boards · Unifix cubes · Dice · Paper plates (to serve as containers for the 100 cubes)

Partners take turns rolling dice to determine the number of cubes to be placed on their place value boards. The object of the game is to be the first one to reach one hundred.

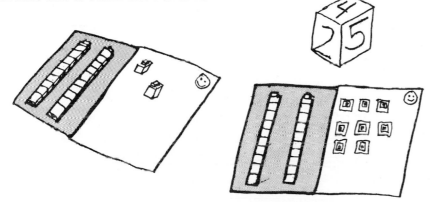

I rolled four. I will have enough for another ten.

Note: There are two possible ways to end the game. A child could need four more to make one hundred and roll a five. That would allow the child to win by getting to one hundred with one left over. It is also possible to establish ahead of time that you have to reach one hundred exactly.

Variation: The winner is the *last* one to reach one hundred.

RACE TO ZERO

Materials: Place value boards · Unifix cubes · Paper plates (to hold the 100 cubes) · Dice

Each partner starts with one hundred cubes and takes away the number of cubes indicated by rolling a die. The winner is the first one to reach zero.

Note: There is only one way to win in this game. An exact number must be rolled to go out. (If a child has four cubes on the board and rolls a five, it is impossible to take five away from four; therefore, that player would miss that turn.)

Section III: Addition and Subtraction of Two-Digit Numbers

In order to make addition and subtraction of two-digit numbers simpler for children, many teachers introduce addition and subtraction with no regrouping before introducing them with regrouping. In the long run this makes it harder for children—not easier. It is misleading to introduce children to addition and subtraction of two-digit numbers with problems in which no regrouping is necessary. In a problem such as

$$\begin{array}{r} 23 \\ + 14 \\ \hline \end{array}$$

, the fact that some numbers are called *ones* and some *tens* is irrelevant to getting answers. If children work exclusively with problems like this for any length of time, they conclude that there is really no difference between $\begin{array}{r}23\\+14\\\hline\end{array}$ and $\begin{array}{r}2\\+1\\\hline\end{array}$ and $\begin{array}{r}3\\+4\\\hline\end{array}$, except that the numbers are written closer together. This conclusion proves false as soon as the children are confronted with a problem such as $\begin{array}{r}25\\+19\\\hline\end{array}$

where regrouping is necessary. This confuses children who have been successful for a long time with problems that looked the same but followed different rules. They become insecure and have to ask the teacher for help in determining when regrouping is necessary. Teachers are bombarded with questions like "Is this a carrying one?" and "Do we have to cross out the numbers this time?"

We can avoid the misconceptions if, instead of first introducing problems with no regrouping, we teach problems in which regrouping may or may not be required from the very first experiences. The children then learn to *always* check to see if regrouping is necessary.

Children should deal with both addition and subtraction problems together so they can compare the two processes and make sense of them. Do not wait for children to master addition before introducing subtraction.

Teacher-Directed Activities

The following activities introduce your children to the processes of addition and subtraction using Unifix cubes. Repeat these activities many times, alternating between problems that require regrouping and those that do not.

Teaching addition and subtraction so that children can make sense of the concepts without learning a ritual for getting answers requires a great deal of time. If children are allowed the time they need at this level to develop understanding, they will avoid years of making errors that come from misconceptions and years of spending time learning and relearning rules for getting answers that make no sense to them.

ADDITION OF TWO-DIGIT NUMBERS——Introducing Addition

Materials: Unifix cubes • Place value boards

When introducing your children to two-digit addition, help them see the logic in what is happening rather than presenting a ritual for them to memorize. The following is an example of how you can lead the children through the steps.

Today you are going to put some cubes on your boards; then you are going to get some more cubes and find out how many you have altogether. What do we call it when we find out how many altogether?

Adding.

That's right. You are going to add. I want you to put out twenty-seven cubes on your boards. Who knows a fast way to count out twenty-seven cubes?

Get two tens and seven more ones.

*Yes. Now I want you to get
nineteen more cubes. This time,
put them under your boards so you
won't get them mixed up with
the twenty-seven that you
already have. Who knows a fast
way to count out nineteen?*

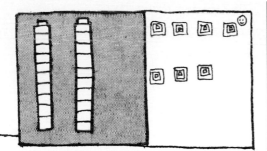

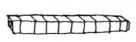

Get one ten and nine ones.

*Now we are going to find out
how many you have altogether.
How many tens and how many
ones do you have when you put
them together?*

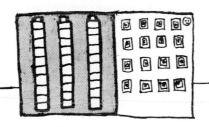

We have three tens and sixteen ones.

*Is that enough to make
another ten?*

Yes.

*Make the ten. How many tens
and how many ones do you
have now?*

Now we have four tens and six ones.

How many is that?   Forty-six.

*Yes, twenty-seven plus nineteen
is forty-six.*

Repeat several times, sometimes presenting problems that require forming tens and sometimes presenting problems that do not require forming tens. The key is *always* asking the children if they have enough ones to make a ten.

Connecting Symbols to Two-Digit Addition

*This time I am going to write down
what we do. Put thirty-four cubes on
your boards. What's the fast way to
do that?*

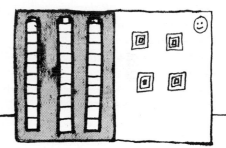

Three tens and four ones.

(The teacher writes **34**.)

*Now put twenty-eight cubes under
your boards. What's a fast way?*

Two tens and eight ones.

(The teacher writes   34
$\underline{+\ 28}$.)

Put them together and tell me how many.

I got five tens and twelve ones.

(The teacher writes
$$\begin{array}{r} 34 \\ + \ 28 \\ \hline 512 \end{array}$$
.)

Is this right? Do you have 512 cubes?

No.

Did you make all the tens you could before I wrote the answer down?

No.

One of the rules for addition is that we have to make all the tens we can before we write the answer. How many tens do you have when you make all that you can?

We have six tens and two ones left.

Now I can write it down.

(The teacher erases **512** and writes
$$\begin{array}{r} 34 \\ + \ 28 \\ \hline 62 \end{array}$$
.)

Let's do another one. Put fifteen cubes on your boards. How can you do that fast?

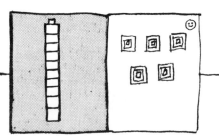

One ten and five ones.

(The teacher writes **15**.)

Put twenty-two under your board.

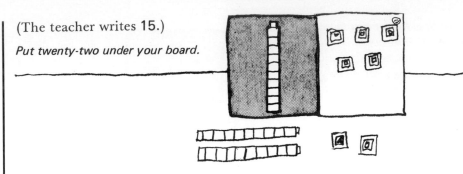

(The teacher writes
$$\begin{array}{r} 15 \\ + \ 22 \end{array}$$
.)

How many do you have when you put them all together?

Three tens and seven ones.

Can you make any more tens?

No.

What do I write?

Thirty-seven.

(The teacher writes
$$\begin{array}{r} 15 \\ + \ 22 \\ \hline 37 \end{array}$$
.)

Now put seventeen cubes on your board and nineteen under your board.

(The teacher writes
$$+\ \frac{17}{19}\ \ .)$$

Put them together and what do you have?

Two tens and sixteen ones.

Should I write it down?

No. We need to make another ten first.

How many tens can you make?

One.

$$\begin{array}{r} 1 \\ 17 \\ + 19 \\ \end{array}$$

I'm going to write that here to show what you did.

How many ones did you have after you made the ten?

Six.

$$\begin{array}{r} 1 \\ 17 \\ + 19 \\ \hline 6 \end{array}$$

I'm going to write that here.

How many tens?

Three.

$$\begin{array}{r} 1 \\ 17 \\ + 19 \\ \hline 36 \end{array}$$

I'm going to write that here.

Lead the children through these same steps on other days until they can tell you what to do and what to write. (You will repeat this lesson many, many times.)

SUBTRACTION OF TWO-DIGIT NUMBERS — Introducing Subtraction

Materials: Unifix cubes • Place value boards

When you introduce subtraction of two-digit numbers, you want the children to think about the process with you and not just memorize a certain sequence of steps. The following lesson example is designed to help you see how you can lead children toward understanding the "rules." You will need to repeat lessons like the following many times. Be willing to discuss over and over with the children the steps that are involved. Understanding the concepts will take time for some of the children but will benefit them greatly in the long run.

Today you are going to take cubes away from the cubes on your boards, and see how many are left.

Put thirty-four cubes on your place value boards. Do you have enough cubes to take twenty-six away?

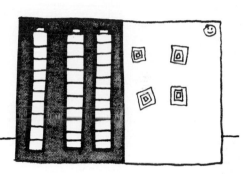

Yes, thirty-four is more than twenty-six.

When you take the twenty-six cubes away, put them under your boards so we can keep track of what you did.

Observe the children as they take the twenty-six cubes away. Some will see immediately that they had to break up a ten to get enough loose cubes. Some will just take four off because they don't have enough loose cubes to take six. Others won't know what to do so will watch their neighbors to see what they do. After everyone has had a chance to think about the process a little bit, discuss what they did.

What did you have to do to take twenty-six cubes away from thirty-four?

Well, the way the cubes were on my board, I couldn't take six off. So I broke one ten up and took two tens and six cubes away, and now I have eight cubes left.

Okay, let's do another one.

*This time put twenty-seven cubes
on your boards. Do you have enough
cubes to take thirteen off?*

Yes.

How many are left?

Fourteen.

Repeat several times, sometimes presenting problems that require breaking up a ten and sometimes presenting problems that do not require breaking tens.

Connecting Symbols to Two-Digit Subtraction

*Now when I give you a problem, I
am going to write it down so we
can keep track of what you're doing.*

(The following procedure will help the children see the *reason* for starting with the ones first when subtracting.)

*Put this number of cubes on your
boards.*

23

*This number tells you how many
to take away.*

$$\begin{array}{r} 23 \\ -\ 17 \\ \hline \end{array}$$

Can we take away one ten?

Yes.

*Okay, take away one ten and tell
me what to write.*

We have one ten left, so write a one.

*Now take away the seven from the
ones side.*

We can't.
Yes we can, but we have to break one up.

*Is what I wrote right? Do you have
sixteen cubes on your boards?
What happened?*

$$\begin{array}{r} 23 \\ -\ 17 \\ \hline 16 \end{array}$$

So what do I have to do?

*Let's do another one. Let's start with
the ones first so we won't have to do
any erasing.*

Put thirty-six on your boards.

$$\begin{array}{r} 36 \\ -\ 19 \\ \hline \end{array}$$

*You want to take nineteen away.
Do you have enough ones?*

*Okay; this time I am going to keep
track of what you are doing. When
you break up one of the tens, do you
still have three tens?*

(The teacher writes
$$\begin{array}{r} 2 \\ 3\!\!\!/6 \\ -\ 19 \\ \hline \end{array}$$.)

Do you have six ones?

No, we have sixteen ones.

(The teacher writes
$$\begin{array}{r} 2\ 16 \\ 3\!\!\!/6 \\ -\ 19 \\ \hline \end{array}$$.)

*How many do you have left when you
take the nineteen cubes away?*

We have six left.
We wrote down the ten left but then
we needed to break it up.

Erase the one ten 'cause we
don't have it.

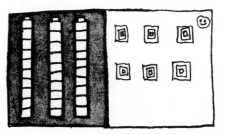

We have to break one up.

No, we have two tens.

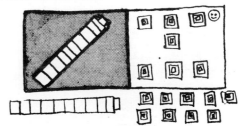

Seventeen.

(The teacher writes
$$\begin{array}{r} 2\,16 \\ 3\!\!\!/6 \\ -\ 19 \\ \hline 17 \end{array}$$
.)

Continue to present a variety of problems for the next several days. Have the children lead you through the steps. Try starting with the tens again so those who missed the point the first time have another chance to see the reason for starting with the ones. The main goal is to keep the children thinking with you. Include some problems that are not possible, such as 34 – 68 or 29 – 42. As children declare they are not possible, cross them out.

MIXING THE PROCESSES

Materials: Unifix cubes • Place value boards

Even before the children have mastered addition or subtraction, present *both* addition and subtraction problems for them to do. It is essential that they learn to switch from one process to the other without needing to be told which is which. Work the problems out as a group so the children can help each other as they learn.

WRITING ADDITION AND SUBTRACTION PROBLEMS

Materials: Unifix cubes • Place value boards • Paper and pencil or Individual chalkboards, chalk, erasers

Once the children are comfortable with the processes, have them write the problems:

Level 1: Provide a model to which the children can refer by writing along with them as they work out the problem with the cubes.

Level Two: The children write the problem as they work out the problem with the cubes. Then you write the correct problem so they can check their work.

WORD PROBLEMS

Materials: Unifix cubes • Place value boards

In the earlier chapter on beginning addition and subtraction, the Unifix cubes were used to represent the objects, people, or animals in the story problems. Because people, objects, and animals do not naturally group themselves into tens and ones, the Unifix cubes are now used as tools for solving the problems rather than as representations of reality. Using the cubes as tools means the children should already be comfortable with adding and subtracting with the cubes. The focus now is on using the cubes correctly to do either addition or subtraction, depending on what the word problem indicates.

For example:

Thirty-five children were riding on a school bus. How can you show that many on your place value boards?

We can put three tens and five ones on our boards.

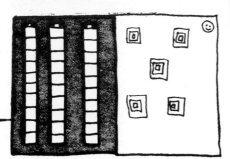

(The teacher writes **35**.)

Four children got off the bus. Do we need to write a plus sign or a minus sign?

A minus sign.

(The teacher writes
$$\begin{array}{r} 35 \\ -\quad 4 \\ \hline \end{array}$$
.)

How many children were left on the bus?

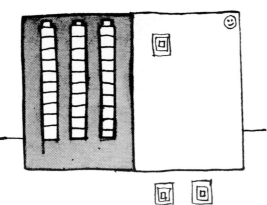

Thirty-one.

(The teacher writes
$$\begin{array}{r} 35 \\ -\quad 4 \\ \hline 31 \end{array}$$
.)

Forty-five people were watching the football game. How can you use your cubes to show that many?

Put four tens and five ones on the boards.

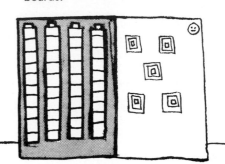

Sixteen more people came to watch the game. What do I need to write?

You need to write *plus sixteen.*

(The teacher writes
$$\begin{array}{r} 45 \\ +\quad 16 \\ \hline \end{array}$$
.)

How many people are watching the game now?

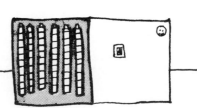

Sixty-one.

Extension: The teacher tells a story as the children set up the problems, using the cubes, and write the appropriate equations.

For example, the teacher says:

John's mother bought a box of oranges. There were fifty-two oranges in the box. She took them to school to share with our class. Twenty-nine children each took an orange. How many oranges did John's mother have left?

The children put out the appropriate number of cubes, write the equation, and figure out the answer.

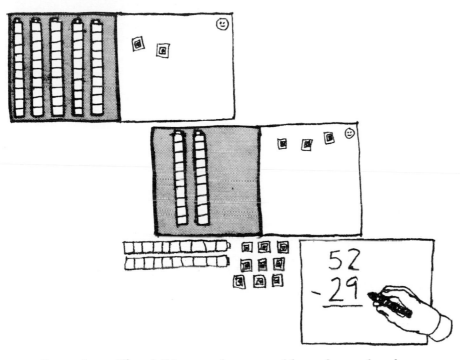

Extension: The children make up problems for each other to solve.

EXPLORING SUMS FROM TEN TO NINETEEN

Because large numbers are hard to visualize, the ability to work easily with sums above ten depends to a great degree on the sense of number the children have developed for the smaller numbers. It is important to help children develop sensible ways of thinking about these numbers so they are not solely dependent on rote memory or laborious counting one by one.

The following activities develop sensible thinking by helping children see how the combinations are related to each other (e.g., $7 + 7 = 14$, so $7 + 8 = 15$). The activities help children focus on the manageable units of tens and ones, thus organizing their thinking.

Related Combinations—How Many Now?

Materials: Unifix cubes · Margarine tubs

Help children use what they know about the combinations for small numbers to figure out answers to sums above ten.

For example:

There are two cubes on this tub and two cubes under the tub.

(The teacher shows the cubes and then writes $\frac{\begin{array}{r}2\\+\ 2\end{array}}{}$.)

How many now?

Four.

(The teacher writes **4**.)

I put one more cube under the tub. How many now?

Five.

Let's write it. How many on top?   Two.

(The teacher writes 2.)

How many underneath?   Three.

(The teacher writes + 3.)

How many altogether?   Five.

(The teacher writes 5.)

Two plus two equals four, so two
plus three equals what?   Five.

Let's do another one.

There are eight cubes on the tub
and eight cubes under the tub.

(The teacher writes + 8 .)
(with 8 above, as $\begin{array}{r} 8 \\ + 8 \\ \hline \end{array}$)

How many now?   Sixteen.

(The teacher writes 16.)

I put one more cube under the tub.
How many now?   Seventeen.

Let's write it. How many on top?   Eight.

(The teacher writes 8.)

How many underneath?   Nine.

(The teacher writes + 9.)

How many altogether?   Seventeen.

Eight plus eight equals sixteen,
so eight plus nine equals what?   Seventeen.

(The teacher writes 17.)

Repeat, using many combinations.

Related Combinations—Stacks

Materials: Unifix cubes

Continue to explore related combinations, this time using stacks of cubes.

   For example:

There are two stacks of cubes: Seven
and seven.

(The teacher writes 7 + 7.)

How many cubes are there
altogether?   Fourteen.

Snap one cube to one of the stacks.

How many now?

                                                    Fifteen.

Let's write it. What do we have?   Seven and eight.

(The teacher writes 7 + 8.)

How many altogether?   Fifteen.

(The teacher writes = 15.)

Repeat many times, exploring a variety of related combinations.

WORKING WITH TENS AND ONES—
Starting with a Particular Number

Materials: "Tens" number shape—see p. 212 for directions for making • Dice

Have the children fill their "ten" shapes with a designated number of cubes. They then roll the dice to determine what number to add to the number they started with.

For example:

Put eight cubes in your "ten" shape.

(The teacher writes **8**.)

Donny rolled a five. Get five cubes.

(The teacher writes $\begin{array}{r} 8 \\ +\ 5 \\ \hline \end{array}$.)

Fill up the ten shape.

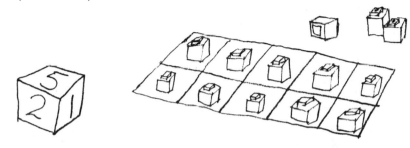

How many did you use?	We used two.
How many left?	Three.
What number did we make?	Thirteen.

(The teacher writes $\begin{array}{r} 8 \\ +\ 5 \\ \hline 13 \end{array}$.)

This is a natural place to discuss informally with the children how the "ten" shape helps them count their answer quickly. After several experiences, some children may grasp the idea of using tens to organize their thinking.

Continue rolling the dice and adding various numbers to the eight cubes. On succeeding days, begin with a different number.

Starting with any Number

Materials: Unifix cubes • Place value board • "Ten" shape • Dice

Have the children take turns rolling the 4–9 dice to determine both numbers to be added, using the "ten" shape.

For example:

Paul rolled a four. Put four cubes in your shape.

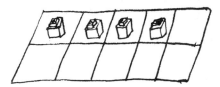

(The teacher writes **4**.)

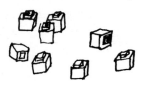

Luanne rolled an eight. Get eight cubes.

(The teacher writes $\begin{array}{r} 4 \\ +\ 8 \\ \hline \end{array}$.)

How many cubes do you think will be left after you fill up the "ten" shape?

I think there will be none left over.

Two.

Let's check and see. Yes, there are two left over. Eight plus four equals twelve.

(The teacher writes $\begin{array}{r} 4 \\ +\ 8 \\ \hline 12 \end{array}$.)

Lynn rolled a seven. Put seven cubes in your shape.

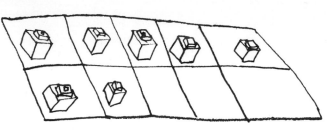

Martha rolled a five. How many leftovers do you think you'll have?

Check and see.

We need three more to fill the shape so we'll have two left over.

Repeat many times.

BUILDING VISUAL IMAGES

Materials: Unifix cubes • Place value boards

After the children have had *many* experiences using the cubes to add and subtract, ask them to try and imagine the board and cubes and figure out their answers in their heads. They should then use the real cubes and boards to check their answers.

Using imaginary cubes is a powerful tool for some children. Others will not be able to visualize the cubes, and their confidence in their ability to make sense of these problems can be damaged. Treat the activity lightly, and don't require a child to do problems without the cubes if he or she is having difficulty.

For example:

(The teacher writes $\begin{array}{r} 23 \\ +\ 16 \\ \hline \end{array}$ on the chalkboard.)

Pretend you are putting two tens and three ones on your boards. Can you see the boards and cubes in your heads? What color cubes are you using?

Green.

Mine are yellow.

Now put one ten and six ones under your boards. We want to find out how many altogether. What do you need to do first?

We should put the ones together.

How many do you have? Remember there's three on your boards and six under your boards.

Nine; we don't have enough to make a ten.

Now what do you do?

We put the tens together. That's two tens and one ten—three altogether.

How many tens and ones?

Three tens and nine ones. Thirty-nine.

Now let's do it with the cubes to see if we get the same answer.

Independent Activities

The children need lots of practice with both addition and subtraction problems. Encourage them to use materials. If some of the children feel they can do the problems without materials, make sure they can prove the answers to you using the materials in their explanation of what the symbols mean. Do not accept descriptions about what to do with the *numbers* (like "cross out the two" or "carry the one"). The children must explain the processes in terms of what happens when they use actual materials.

For example:

Tell me how you got this answer.

$$\begin{array}{r} {}^{3}\!\!\!{}_{1} \\ 4\,2 \\ -\ 2\,4 \\ \hline 1\,8 \end{array}$$

I knew I couldn't take four away from two, so I had to break up a ten and put it with the two, and that made twelve. That meant I only had three tens left. I took four away from twelve, and that was eight. I took two tens away from three tens and that left one ten.

PARTNER ADD-IT

Materials. Unifix cubes • Place value boards • Paper and pencil

Partner A puts some cubes on the place value board, and Partner B places cubes under the board. They each write the problem created, after which they put the cubes together and write the answer.

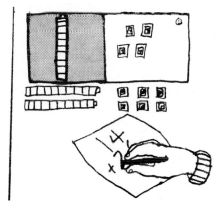

Nancy:

I put one ten and four ones on the board.

Julie:

I put two tens and six ones under the board.

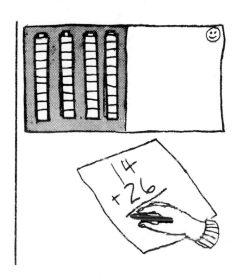

We can make a ten.

That makes four tens and no ones.

Extension: After the partners have placed the cubes on and under the board and have written the problem, they try to figure out the number of cubes altogether. They then check their answer by actually putting the cubes together.

PARTNER TAKE-AWAY

Materials: Unifix cubes • Place value boards • Paper and pencil

Partner A builds a number on the place value board. Both partners write the number on a piece of paper. Partner B takes some of the cubes away (putting them under the board rather than back with the other cubes). The partners write the number taken away then see how many are left and write that number down.

Mike: I put four tens and two ones on the board.

Ramon: I took one ten and five ones. First I had to break one of the tens up because I wanted to take off more than two ones.

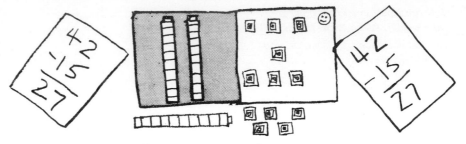

There are two tens and seven ones left on the board.

Extension: Have the child who takes the cubes away hide the number left by placing a sheet of paper on the board. The partners try to figure out the number left then lift the paper to check their answers.

WORD PROBLEMS

You and the other children can write word problems for the children to solve using the cubes.

On Monday 16 children at Maple School had chicken pox. On Tuesday 12 more got chicken pox. How many children in Maple School have chicken pox?

Extensions:

Mail-Order Catalogs
After instruction in writing dollars and cents, have the children use a mail-order catalog to choose things they would like to buy. They can use the cubes to figure out how much things cost.

Newspapers
Have the children find things that they would like to buy in newspaper advertisements. Use the cubes to help determine how much they spent.

Menus
Have the children choose what they would like to order from a restaurant and add up the price of the meal, using the cubes to help them add.

FIGURE IT OUT

When children are taught a procedure for doing something (like adding and subtracting using the place value boards), they often get locked into following that procedure. To get them to realize that sometimes they can figure out the answers without actually going through the learned ritual, present the following types of problems. Discuss each type of problem with the children, and help them see the patterns and relationships.

For example:

24 + 1	36 + 1	43 + 1	98 + 1
23 - 1	22 - 1	19 - 1	64 - 1
35 + 2	43 + 2	16 + 2	38 + 2
9 + 5	29 + 5	49 + 5	69 + 5
5 + 3	25 + 3	45 + 3	95 + 3

ROLL AND ADD—ROLL AND SUBTRACT

Materials: Tens dice (labeled 0 tens–5 tens) • Ones dice (4–9) • Place value boards • Unifix cubes • Adding or subtracting worksheets that are cut into strips—see black line masters 73, 74 (These worksheets can be used over and over again because new problems will be created each time the children roll the dice.)

Children create a variety of addition and subtraction problems using special dice and open-ended worksheets. They do as many worksheets as time allows, selecting adding or subtracting strips.

David: My tens dice has three and my ones dice has six.

I can write it under the eighty-six and work the problem.

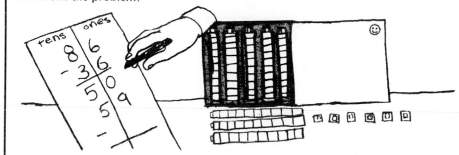

Now I'll roll again to see what I have to take away from 59.

Carl: I'm adding this time. I rolled nine ones and two tens.

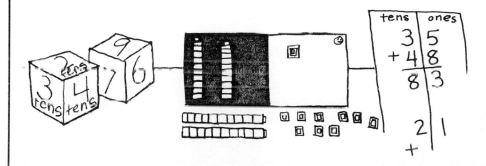

I almost put the two in the wrong place. It's two *tens*.

TEXTBOOK PAGES AND WORKSHEETS

Once children understand the processes of addition and subtraction, they can work problems in the textbook for practice, using the cubes to work out the problems. Children should be required to use the materials until they can explain the problems to you in terms of what would happen if they used the cubes.

ANALYZING AND ASSESSING YOUR CHILDREN'S NEEDS

The concept of place value is not an easy one, especially for young children. Understanding develops and broadens with many experiences over a long period of time. The activities in this chapter, designed to help children *begin* to understand, provide a solid foundation that will serve students well as they move on in school.

When children are first exposed to place value activities, their focus is often on what to do. They are concerned about such things as "Where did the teacher say to put the tens?" or "What number am I supposed to write down?" Over time, as the procedures involved in working with the materials become familiar, children are then able to think about the concepts, form generalizations, and increase their understanding of place value. It is normal that they will be confused during this period. Many children will appear to understand in some settings and then appear not to understand in other settings. The following kinds of behavior are not unusual. They can serve as clues to children's levels of thinking, and they can indicate the need for continuing to work with the place value ideas.

Patrick has counted cubes one by one and has filled a container. "I have thirty-five cubes in this bowl," he says. "How many tens do you

think you can make with all those cubes?" his teacher asks. "I don't know," Patrick says. "Maybe about six." "Try it and see," his teacher says.

Patrick does not yet see the connection between the number *thirty* and the quantity of three tens, so knowing he has thirty-five cubes does not give him any clue about how many tens he has.

◆ ◆ ◆

When given the written clue *26,* Michael put two cubes on the blue side of his board and six cubes on the other.

Michael is still responding to the numeral *2* the way he has for a long time—as representing two objects. It is a big step for a child to learn that the numeral can stand for "two groups of ten." Sometimes Michael does remember that, but, because he is still in the learning stages, he sometimes forgets.

◆ ◆ ◆

Jody's teacher has placed a pile of cubes on the table. She asks Jody to make all the tens she can. Jody snaps together cubes into groups of ten. She has three tens and six left over.

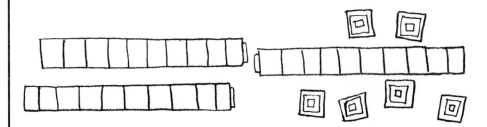

"How many cubes do you think are on the table?" asks Jody's teacher. Jody begins to count each cube one by one until she counts them all.

The same task is given to Ginnie. She arranges the cubes into piles of ten. When asked to tell how many, she immediately says, "Thirty-six."

Jody knows she has three tens and six ones, but grouping cubes into tens and ones does not give Jody the information she needs to determine she has thirty-six altogether. She needs to count each cube before she is sure of the amount. Unlike Jody, Ginnie has discovered the relationship between three tens and six and the quantity *thirty-six,* and she knows how many cubes are there without counting by ones.

Becky has counted out twenty-four cubes one at a time. Her teacher then asks her to put them into tens and ones. She asks Becky, "How many cubes do you have now?" Becky recounts all the cubes one at a time to determine that she has twenty-four.

Although Becky knows she started with twenty-four cubes, she is still not sure that rearranging cubes into tens and ones has not changed the quantity. She does not have conservation of large numbers.

The teacher says, "Show me fourteen." Gina counts out fourteen cubes and places them all on the ones side of her place value board. After she clears her board, the teacher then asks her to show "one ten and six." Gina puts one ten on the tens side of her board and six cubes on the ones side. "Now show me twenty-one," her teacher says. Gina counts out twenty-one cubes and puts them on the ones side of her board.

When Gina hears a number such as fourteen or twenty-one, she does not yet think of them in terms of tens and ones. Rather, she thinks of a total group of the particular quantity. Only when the teacher says "two tens and one" or "one ten and four" does she group them that way.

Jorge's teacher asks him, "How many tens do you have on your board?" He says, "I have thirty tens."

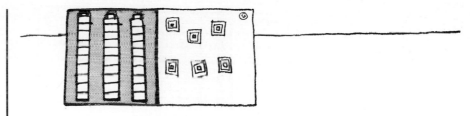

Barbara's teacher asks her how many cubes she has on her board.

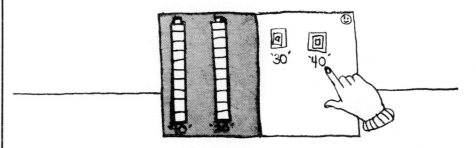

Barbara begins to count by tens and says, "Ten, twenty, thirty, forty. Here's forty cubes."

Barbara and Jorge are still not used to speaking in the language of tens and ones. They are not having problems with the concepts; they just didn't say what they meant to say.

The above-mentioned children are thinking in ways that are very normal for children just learning about place value. They simply need many more experiences grouping large numbers into tens. In time, these concepts will become clear to them.

Carol has been assigned some subtraction problems. She is supposed to use the Unifix cubes to solve them. The first problem she is to do is

$$\begin{array}{r} 23 \\ -\ 17 \\ \hline \end{array}$$

She puts out the correct number of cubes to begin with.

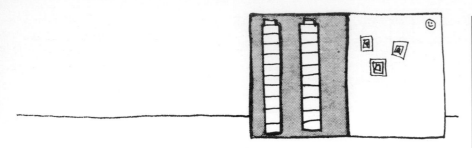

Carol then begins to ponder how she is going to take away seven. She reaches over her board to the extra pile of cubes, gets four more cubes, and puts them with the other three. Now she has enough to take away seven.

I need to get more so I can take away 7.

She then takes away the seven cubes, takes away the one ten, and writes her answer.

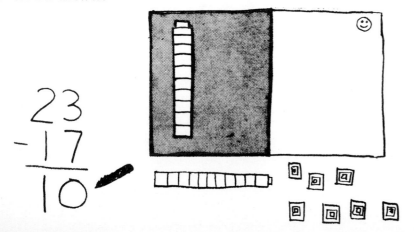

$$\begin{array}{r} 23 \\ -17 \\ \hline 10 \end{array}$$

Carol solved the problem in a way that made sense to her but that shows she does not realize what is allowed when subtracting. She remembers the teacher said you could get more when you were subtracting, but she doesn't yet know it has to be taken from the "tens" and not from the table. She should not be working with subtraction independently; she needs more instruction from the teacher.

◆ ◆ ◆

Jenny has demonstrated for her teacher her ability to do "carrying."

$$\begin{array}{r} 1 \\ 26 \\ + 39 \\ \hline 65 \end{array}$$

"Six and nine is fifteen. Put down the five and carry the one. One and two is three and three more makes six." Her teacher says, "That's good, Jenny. Now show me that with the cubes." Jenny puts out two tens and six and then three tens and nine, and she pushes them together.

Jenny looks at her results and says, "Oh, yeah. I made a mistake." She then erases her original answer and writes

$$\begin{array}{r} 1 \\ 26 \\ + 39 \\ \hline 515 \end{array}$$

Jenny has learned to work with symbols without understanding. Her lack of understanding shows up when she works with the cubes.

◆ ◆ ◆

Rick has been given the problem

$$\begin{array}{r} 24 \\ -\ 16 \\ \hline \end{array}$$

He puts out Unifix cubes in the following manner.

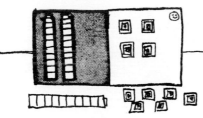

Then he says, "I don't know what to do now."

Rick sees the two symbols and immediately thinks of two sets. He needs to be reminded of what subtraction means; that is, that the second number is not another set but a clue that tells him what to take away. Rick has been successful doing subtraction problems in the past, but, because he is still learning and has not mastered it yet, he will continue to be confused on occasion.

◆ ◆ ◆

Even after many lessons dealing with place value, children will make mistakes like those just described. When teachers see that using materials to teach children does not prevent these mistakes, some stop using materials and begin teaching the children to work only with symbols. They believe the materials are confusing the children. The reality is that the materials show a child's confusion, which is often masked by the ability to get right answers when given symbols. Rather than taking the materials away, the teacher should recognize that the confusion is an indication that the child needs more experiences with the models.

Children should be allowed to work with symbols without materials only after they have proven to you their ability to show addition and subtraction problems using the models. They should be able to write the symbols that describe the processes they have done with the cubes, and they should be able to explain what the symbols represent in real-world terms. If they cannot do these things, then continue to provide a variety of experiences with materials until they have a good understanding of place value.

CHAPTER SIX

If your workbook or textbook objectives
are   ● Introduction to multiplication
      ● Multiplication or single-digit numbers

Then you are dealing with

Beginning
Multiplication

WHAT YOU NEED TO KNOW ABOUT BEGINNING MULTIPLICATION

Knowing how to do "times" is for many children an entry into the world of the big kids. Many get parents or big brothers and sisters to teach them the times tables. They come to school filled with pride, eager to show the teacher what they can do. Many of the children who can rattle off such phrases as "ten times ten is one hundred" or "two times three is six" have no idea what number relationships they are expressing. They do not understand that those numbers they are saying mean something in the real world.

One teacher asked a child to show him "four times two" with the cubes. The child did this:

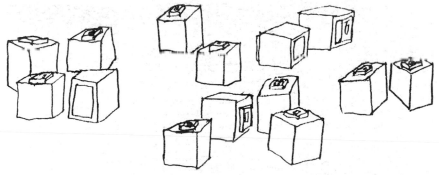

I know the answer. The answer is eight!

Although knowing answers may be sufficient to satisfy children's needs to be able to do what their big brothers or sisters can do, it is not a legitimate goal for teachers of young children. Knowing answers to multiplication problems is useful at some stage, but it is useful only if that knowledge is based on meaning and understanding.

Your goal should be that children become familiar with the process of multiplication as it occurs in the real world. Children should learn to write the symbols used to describe these processes only after they can interpret the oral language patterns easily.

Section I: Developing the Concept of Multiplication

Teacher-Directed Activities

Because your first and most important goal is to help children develop the language of multiplication, you will need to provide a variety of experiences that can be described with natural language from the real world. Terms such as *rows of*, *stacks of*, and *groups of* will be used in the following activities. It is very important not to introduce the formal term *times* too soon. This word interferes with many children's ability to visualize groups (and for those who have been exposed to the word elsewhere, it often creates misconceptions).

The following activities should be repeated many times so that children can become flexible and feel at ease with any description of multiplication. Some children will have difficulty interpreting the language used, because the process of multiplication requires them to think about and count groups of objects rather than single objects. To help these children begin to relate the language to what it represents, build models using the Unifix cubes along with them when giving directions until they can build the models themselves.

LOOKING FOR EQUAL GROUPS IN THE REAL WORLD

An important concept children must understand to be successful with multiplication is the idea that multiplication is counting groups of equal number. Help the children find equal groups by taking walks or looking around the classroom.

Looking for Groups of Equal Number in the Classroom

For example:

I see six panes in our window. Are there any other windows in our room that have six panes?

I see two more.

Yes, we have three windows with six panes. Six panes and six panes and six panes. Today we want to look for other things that have the same number of things over and over, like the windows.

Look at our boxes of watercolors. Do all these boxes have the same number of colors in them?

Yes. There's eight and eight and eight.

Yes, each box has eight colors. We have eight here and here and here and here and here and here. How many boxes of paint is that?

Six.

Yes, we have six boxes of eight colors.

We have some tables in here that all have six chairs at them. How many tables have six chairs?

There are four tables that have six chairs.

Yes, we have four tables with six chairs. There are six and six and six and six.

Looking for Equal Groups Outside the Classroom

Let's go for a walk and see what we can find that has equal groups (that means something with the same number over and over and over).

Here are three cars parked altogether. Do they each have something that's the same number as the other cars?

I see the lights. This car has two lights and this car has four lights. That's two lights and four lights.

We are not looking just for things that look the same. We are looking for things that are the same number. Are two and four the same number?

No, but I see something that's the same number. The wheels are the same. See, four and four and four.

Yes. That's right. That's three groups of four.

Extension: After many experiences looking for groups—when the children have a clear understanding of what it means to be looking for things with the same number—they can use the Unifix cubes in the classroom to represent the things they saw on their walks.

For example:

We saw three cars with four wheels. Can you show me the sets of wheels? Don't show me the whole car; just the wheels.

We saw five streetlights with two lights on each pole. I can show that with my cubes, too.

ACTING OUT MULTIPLICATION STORIES—Using Real Things

Materials: Things readily available in the room

Tell the children stories, and have them act out the stories using real things readily available in the room. Have the group direct the children who are acting out the story so that no individual is put on the spot.

For example:

Paul and Linda line up some chairs. They make three rows with four chairs in each row. How many chairs do they line up?

Dennis, Frances, Kathy, Bernadette, and Jamie put five chairs at three tables. How many chairs do they use?

Manuel gives five children two pencils each. How many pencils does he pass out?

Bonnie makes four stacks of books. She puts three books in each stack. How many books does she use?

Carolyn puts five boxes of crayons on the table. Each box holds eight crayons. How many crayons are in the boxes?

Lee puts three erasers at each table in the room. There are six tables. How many erasers does he put out?

BUILDING MODELS OF MULTIPLICATION PROBLEMS

Materials: Unifix cubes

The following activities require the children to interpret various language patterns and build the appropriate models. This lesson will need to be repeated over and over again for some children. Build the models along with the children while they are learning.

Stacks

Build three stacks of five.

The children build:

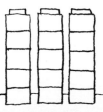

Build two stacks of four.

Build four stacks of six.

Rows

Make three rows of four.   The children build:

Make two rows of three.

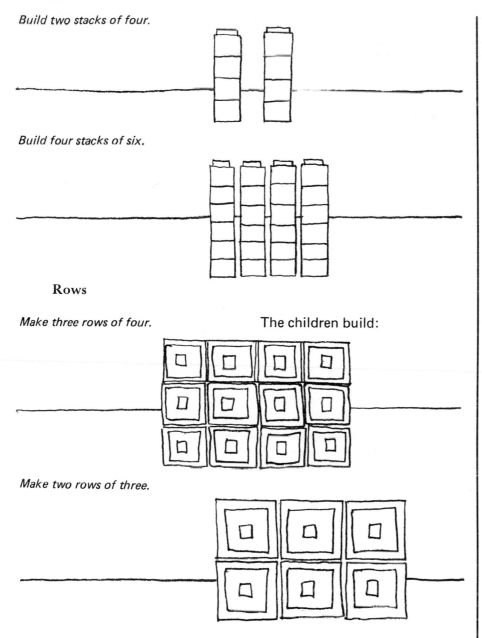

Note: Define a row as going in the direction that your arms are in when you hold them out to the sides.

Ask the children to make the rows touch. This will help prepare them for later work with arrays and area.

Groups   The children build:

Make two groups of five.

Make three groups of four.

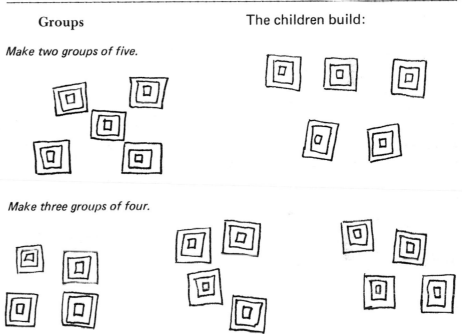

As was discussed earlier in the place value chapter, children must think differently when counting groups of objects from counting single objects. The following procedure can help straighten out in children's minds which number refers to rows or groups and which number refers to the number in each row or group.

Start with one row (or stack or group), and add additional rows one at a time.

For example:

This is one row of five. You build one row of five.

How many rows have you built?   One.

Yes, you built one row of five.

Here is another row of five. Now I have two rows of five. You build another row of five. How many rows of five do you have now?

Two rows of five.

As soon as the children are confident, change the number of cubes in the row.

For example:

Now what do I have?   Three.

Yes, I have one row of three. Now what do I have?   Two rows of three.

Yes. Now you build four rows of three, seven rows of three, etc.

If the children are successful, give a variety of directions, sometimes changing the number of rows and sometimes changing the number of cubes in each row. If they appear confused or hesitant, ask the children to build one row (or stack or group) again, and help them straighten out their thinking again.

BUILDING RELATED MODELS

Materials: Unifix cubes

Once children are able to build easily rows, stacks, or groups according to your direction, present related facts such as six groups of two and two groups of six.

For example:

Build two rows of three.
Build three rows of two.

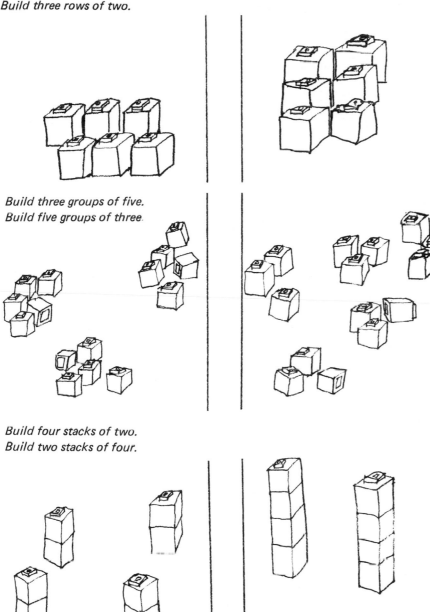

Build three groups of five.
Build five groups of three.

Build four stacks of two.
Build two stacks of four.

Note: It will not be obvious to all children that equations such as these require the same number of cubes. Do not take away the children's opportunity to discover this for themselves. Your responsibility is to provide situations in which those relationships are evident. Children should have the opportunity to discover these relationships in their own time. Simply observe the way they solve the problems at this stage of their development.

ACTING OUT MULTIPLICATION STORIES—Using Unifix Cubes

Materials: Unifix cubes • Working space papers (non-dotted side)

Tell the children stories, and have them act out the stories using Unifix cubes to represent various people, animals, and objects. When the children listen to these stories, they must be able to pick out the appropriate objects to be built and leave out the inappropriate objects.

For example:

There are four houses on Letitia's street. Each family has two cars. How many cars do they have?

What you are asking about is the number of cars—not the number of houses. The houses is the clue to the number of groups. This will be confusing for some children, so talk the problem through with them, and build the model for them to copy. It can help if you say, "I want to see only the cars. How many cars are in front of the first house? Show them with the cubes. How many cars are in front of the second house? Show those with the cubes. Is that all the houses? Show me the cars in front of the next house. Show me the cars in front of the next house. We made four groups of two cars. How many cars in all?"

Note: If some children need to see the houses, have them use a Unifix cube to represent each house, then have them put two cars in front of each house. When you ask the question "How many cars?" make sure they count only the cars and not the houses.

Examples of story problems:

Tim has three dogs. He gave each dog two bones. How many bones did he give his dogs altogether?

Five kids went to the library. They each checked out three books. How many books did they check out altogether?

There are five people in Dale's family. Each person gets to put four decorations on the Christmas tree. How many decorations do they hang?

Robin's mother went shopping for school clothes for the three children in Robin's family. She bought three new shirts for each child. How many shirts did she buy?

Section II: Connecting Symbols to the Concept of Multiplication

Teacher-Directed Activities

Too often children learn multiplication tables purely by rote with no understanding of what they are memorizing. The activities in this section are designed to help children make the connection between the symbols and what they stand for.

MODELING THE RECORDING OF
MULTIPLICATION EXPERIENCES

Materials: Unifix cubes

When the children can interpret a variety of language patterns representing multiplication, begin writing symbols to demonstrate for them the way these experiences can be written down. It is critical that the symbols used connect in the child's mind to the experiences they have been having with multiplication. To help make that connection, begin recording the experiences using the written words rather than the times sign. Be very careful not to use the word *times*.

Misty stacked up books into two piles.
She put four in each pile.

(As the problem is being acted out, say and write the following.)

How many stacks is Misty making?   Two.

(The teacher writes **2 stacks of.**)

How many books are in each pile?   Four.

(The teacher writes **2 stacks of 4.**)

How many books altogether?   Eight.

(The teacher writes **2 stacks of 4 = 8.**)

Mark lined up two rows of four chairs.
How many rows did Mark make?

(The teacher writes **2 rows of.**)

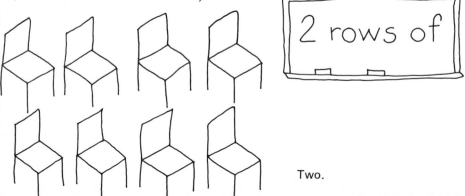

Two.

How many chairs in each row?　　Four.

(The teacher writes **2 rows of 4**.)

How many chairs did he use?　　Eight.

(The teacher writes **2 rows of 4 = 8**.)

Brian is choosing people to be on two teams. He needs six people on each team. How many people does he need on both teams?
How many teams are there?　　Two.

(The teacher writes **2 teams of**.)

How many on each team?　　Six.

(The teacher writes **2 teams of 6**.)

How many kids altogether?　　Twelve.

(The teacher writes **2 teams of 6 = 12**.)

INTRODUCING THE MULTIPLICATION SIGN

Give word problems, and have the children act them out as before. As the story is being acted out, write the equation, using words, as described earlier.

　　For example:

Mary's father was making pancakes for three people. He stacked four pancakes on each of their plates. How many pancakes did he make?

(The teacher writes　3 stacks of 4 = 12)

　　Tell the children that there is an easier way to write the word problem. Erase the words *stacks of* and substitute the times sign. (Say "stacks of" as you write the sign.)

3 stacks　　3 X

Note: Some children will recognize the sign and call it "times." Acknowledge that it is the times sign, but say that it means "stacks of" in this sentence. Do not allow the children to read the word as *times*, but do not say it is incorrect; just tell them that they need to learn to read it with the words. Repeat the activity, telling more word problems with various vocabulary (*groups of, rows of,* etc.). Each time replace the words with the times sign. Tell the children the sign can be read as rows, stacks, or groups.

INTERPRETING SYMBOLS

Put multiplication problems on the chalkboard, and ask the children to act them out using the cubes.

　　At first, have one of the children choose the way they will read the times sign (as *rows of, stacks of,* or *groups of*). Then all the other children will read it that way and build the appropriate model.

　　5 X 3

　　For example:

　　　　　　　　　Let's make stacks this time.

How many stacks do you need to make? How many in each stack?

　　Extension: Let each child choose her or his own way of interpreting the problem and share it with the others.

　　For example:

　　4 X 3 = 12

Four rows of three is twelve.

CUPSFUL

Materials: Unifix cubes • Plastic cups • Dice (optional) • Worksheets—see black-line master 76 (cut in half, use half shown in drawing below)

The child takes the number of cups she or he wants to use for the activity (or the child can roll a die to determine the number of cups). He or she writes the number of cups in the appropriate place on the worksheet, then uses the cubes and cups to solve the problems and writes the answers.

For example:

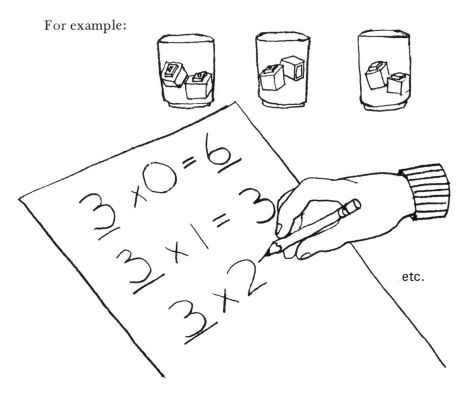

I used four cups yesterday. Today
I want to use three cups.

Extension: Looking for Patterns

The children can record the information on the worksheet and look for patterns.

Have them record the answers on the strip used in the place value activities (see page 145). Have them loop the patterns they find and then extend the pattern without using the cubes.

For example:

Have the children color in the answers on the 00–99 chart, then have them extend the pattern without using the cubes.

For example:

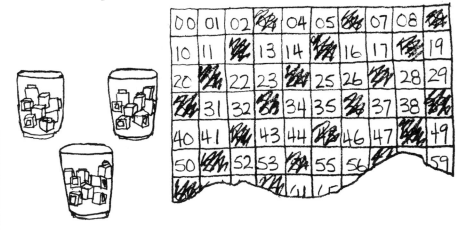

SQUARES AND RECTANGLES

Materials: Unifix cubes • Task cards with various squares and rectangles on them—see examples below • 2 x 6 pieces of paper

The children fill in the various squares and rectangles and write the multiplication equation that describes them.

For example:

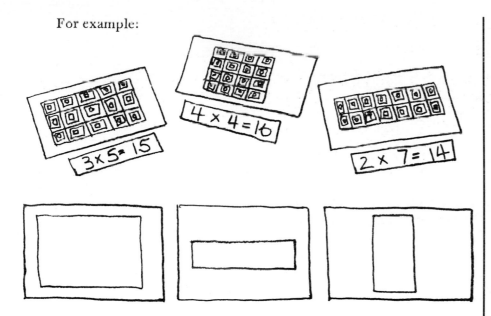

PARTNER PROBLEMS

Materials: Unifix cubes • Working space papers (non-dotted side)—see p. 210
• Paper and pencil or individual chalkboard, chalk, eraser—see p. 215

Partner A makes up a problem for Partner B by building a model. Partner B writes the multiplication equation and answer. Then Partner B makes up a problem for Partner A.

For example:

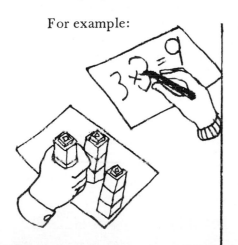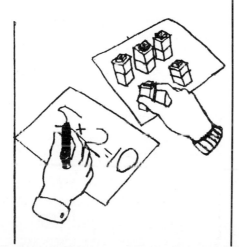

ROLL AND MULTIPLY

Materials: Unifix cubes • Dice • Worksheets—see black-line master 77 (cut in strips)

The following activity allows children to create a variety of multiplication problems using open-ended worksheet strips, Unifix cubes and dice. They roll dice to determine the problems, build the multiplication equations, and record their answers on the worksheets. They do as many worksheet strips as time allows.

For example:

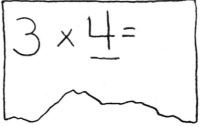

I rolled a four.
I write a four on my worksheet.

Worksheet strip

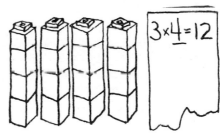

I build three groups of four. That makes twelve. I write twelve on my worksheet.

This time I rolled a two.

The child continues until the worksheet strip is completed.

WORKBOOK PAGES AND DRILL SHEETS

The children can use the cubes to figure out the answers to problems in their workbooks. If some children do not need to build the models in order to write the answers, allow them to stop using the cubes only after they have demonstrated to you that they know how to build the appropriate models.

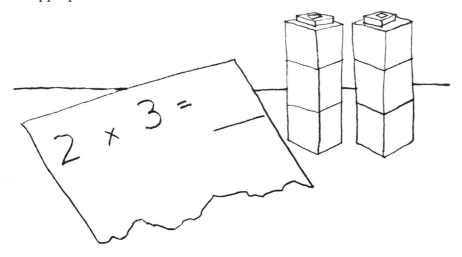

UNIFIX PUZZLES

Materials: Unifix cubes • Unifix puzzles—see p. 211 for directions for making • 2 x 6 pieces of paper

The Unifix puzzles that were used for counting and for addition and subtraction practice can also be used for multiplication, and they can provide a challenge for those children who are ready for one. The puzzles provide an opportunity for children to learn to describe arrangements of cubes as combinations of multiplication and addition and can introduce them to the use of parentheses.

For example:

(a)	(b)

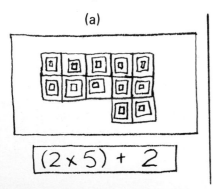

$$(2 \times 5) + 2$$

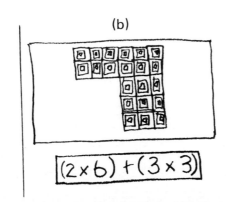

$$(2 \times 6) + (3 \times 3)$$

(c)

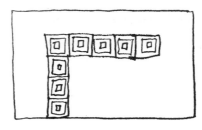

$$(1 \times 5) + 3$$

ANALYZING AND ASSESSING YOUR CHILDREN'S NEEDS

It is important that children have a strong foundation in understanding multiplication before we push them into the rote memorization of basic facts. The following examples will provide you with clues for determining your students' understanding.

INTERPRETING THE LANGUAGE

When Florence is asked to build models such as three stacks of two, she gets mixed up and has trouble deciding if she is to build two threes or three twos. The total number of cubes is the same either way and would give her the correct answer; however, it is important that she learn to interpret the language accurately. Setting up five rows of three chairs is different from setting up three rows of five chairs. Serving three guests five cookies each is not the same as serving five guests three cookies each.

When Rose is asked to build three rows of two, she hears the *three,* the *two,* and the word *rows,* so she lines up two cubes in a row and then three cubes in a row.

Sharon was using the cubes to act out the following story problem.

Simon had four rabbits. He gave each rabbit two carrots. How many carrots did he give them?

Sharon put out cubes in the following manner and called out *twelve* as her answer.

Sharon put out cubes to represent the rabbits as well as the carrots. Her teacher needs to repeat the question, asking Sharon to tell her only how many carrots—not how many cubes she used to act out the story.

The children in the previous examples do not yet understand the language used to describe multiplication situations. Their lack of understanding is evident when they are asked to build models. Their teacher needs to give them opportunities to hear the language and see the related models. When she asks them to build any models, she should build right along with them until they are ready to do the work on their own.

UNDERSTANDING THE PROCESSES

Mike was very proud of the fact that his big brother had taught him to write answers to "times" problems. He had memorized many of the basic multiplication facts. One day, he was asked to fill in a rectangle that took two rows of six cubes.

He was also asked to write an equation describing the rectangle. He wrote *3 x 4 = 12.* The teacher told him that was not what he saw, and Mike said, "There's twelve cubes in the rectangle, and three times four equals twelve."

The teacher asked John to use cubes to show him "two times three." John put out a set of two cubes and a set of three cubes. He said, "I know the answer. It's six."

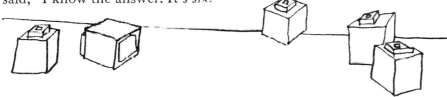

Both boys knew more about the answers than about the real situations from which those answers were supposed to come. They need to learn that the basic facts they know by rote can be repre-

sented logically with cubes. Their teacher should present the lessons that introduce multiplication with story problems and the building of models. The teacher needs to demonstrate for them that these real-life models can be described with those very equations they had previously memorized without understanding.

WORKING WITH SYMBOLS

Clara was working with the counting boards and the multiplication equation cards. She had some answers correct, but for others she added instead of multiplied.

Gerald had just been introduced to division during his directed teaching time. During his independent time, he was supposed to work with multiplication cards and counting boards. He had already worked for several days with them. He sat looking at his equation cards until the teacher looked up to see how those working independently were doing and noticed Gerald wasn't doing anything. "I forgot how," he said.

Clara had worked for a long time with addition before she learned to multiply. It is to be expected that she may forget sometimes and slip back to doing something familiar. Gerald is in the process of learning something new and now is confused about what he learned before. It is a natural part of the learning process to have times of confusion and forgetfulness. Clara and Gerald's teacher needs to accept that fact and simply remind them of the processes, reviewing with them as often as necessary.

The child writes *3* in the appropriate place on his recording sheet.

Now I'll get twenty cubes and put them in three rows on my division card.

Oh, two are left over; they go up on top. They are called remainders.

There are six in each row and two left over. I'll write it down on my recording sheet.

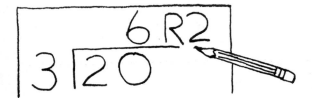

PARTNER PROBLEMS

Materials: Unifix cubes • Paper or individual chalkboards

Partners make up problems and record them.

For example: Partner A picks a number of cubes.

Partner A: This train is fourteen cubes long.

Partner B decides what he or she wants to break the train into.

Partner B: Let's break it into twos.

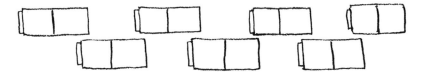

Partner A writes the equation that describes what Partner B did and what the results were.

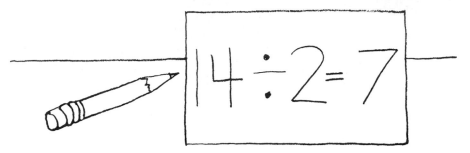

The partners continue to make up problems, switching roles.

CUPS OF CUBES

Materials: Unifix cubes • Plastic cups • Worksheet—see black-line master 79

The children choose the number of cups with which they want to work. They divide various numbers of cubes into those cups and write the equations on worksheet 79.

For example:

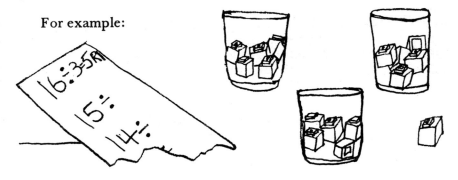

I will work with three cups.

USING THE 100 TRACK FOR DIVISION

Materials: Unifix cubes • Unifix 100 Track (see p. 218) • Division markers (see p. 218) • Paper

The children place a number of cubes in the Unifix 100 Track. They count off by a particular number and place the division markers in the appropriate places.

For example: I put in eighteen cubes.

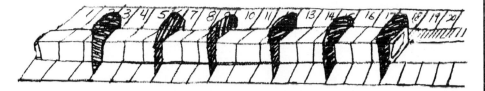

I counted to three and put a three
marker each time I got to three.

The children determine the number of markers used and write the division equation.

$$18 \div 3 = 6$$

Continue the game, changing the number of cubes being worked with.

For example: I put in fourteen cubes.

$$14 \div 3 = 4 \ R2$$

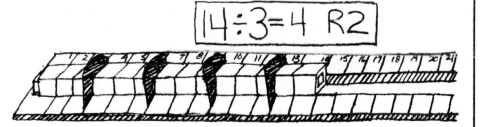

Or changing the number being used as the divisor.

For example:

$$14 \div 4 = 3R2$$

I put in fourteen cubes. I am going
to divide by four.

CREATION TASK CARDS FOR DIVISION

Materials: Unifix cubes • Worksheet—see black-line master 79 (cut apart) • Creation Task Cards—see p. 213 for directions for making

Each child chooses a Creation Task Card. The children are to determine how many of these creations they can make from the number of cubes indicated on their worksheets. They record their answers on their division worksheets.

For example:

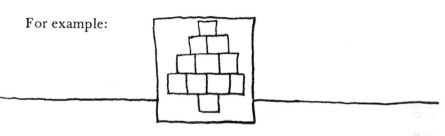

I picked a tree.

It takes eleven cubes.

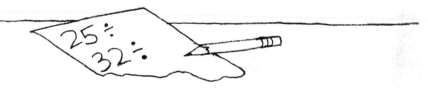

My first number is twenty-five.

How many trees can I make?

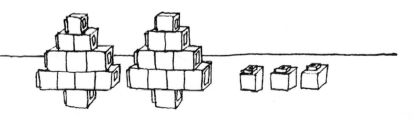

I made two, and three were left over.

(The child writes the equation on her or his worksheet.)

My next number is thirty-two.

Note: Children may do the entire worksheet using only one creation (the same number). Others may choose to pick a different creation for each quantity they divide. This freedom provides many opportunities for practice with a wide variety of problems.

WORKING WITH MULTIPLICATION AND DIVISION TOGETHER: WORD PROBLEMS

When children are comfortable with both multiplication and division, begin presenting opportunities for them to deal with both processes.

The teacher presents multiplication and division word problems. The children use cubes to represent the objects in the stories and write the equations that describe the action.

Tell both kinds of word problems, and have the children distinguish between the two processes.

WORKING WITH ALL FOUR OPERATIONS—Comparing Equations

The following game can be a challenge for two players who are confident with all four processes (addition, subtraction, multiplication, and division).

Mix up equation cards with all four processes on them. Each partner draws one card. The players use the cubes to build what is indicated on their cards. They then turn the more/less spinner to determine if the person with more or less cubes wins the two cards.

For example:

Partner A   Partner B

Less wins. I keep both the cards.

ANALYZING AND ASSESSING YOUR CHILDREN'S NEEDS

The following are examples of problems that children encounter when working with division.

UNDERSTANDING THE PROCESS

David was asked to divide six cubes into three cups. "How many cubes do you have in each?" his teacher asked. "I have two and two and two," David said.

David must learn that he needs to tell only what is in *each* cup—not *every* cup. In this case, saying a number one time refers not to one group but to several groups of the same size.

◆ ◆ ◆

Nancy broke a train of twelve cubes into lengths of four. When asked to tell the results, she said, "There's four in each train." She wrote $12 \div 4 = 4$.

Nancy is focused on the number of cubes in each train rather than the number of trains she could make. Her teacher needs to ask again so Nancy can refocus on what she is trying to find out. Her teacher says, "We are trying to find out how many trains we can make when we break twelve cubes into trains of four. How many trains can you make?" With enough experiences, Nancy will find it easier to describe what she is doing with the cubes.

◆ ◆ ◆

Carlene divided seventeen cubes into four cups. She put four cubes into three of the cups and five cubes into the fourth cup. "I can't do this one," she told her teacher. "It's got two answers."

Carlene's teacher needs to remind her that each cup *must* have the same number. If she can't put the same number in each, she has to leave the extra ones out. The teacher tells her that the extras are called *remainders*.

◆ ◆ ◆

WORKING WITH SYMBOLS

Bill has practiced doing division problems at home and knows how to write answers easily. However, when given a story problem that describes a division situation, he has no idea how to write it down.

Bill knows how to get answers but does not really see the connection between the process he does on paper and the real world. His teacher needs to do some story problems with Bill and the other students in the same situation. They should act out the problems and observe how the teacher writes equations to describe them. Soon they will be able to write equations themselves.

◆ ◆ ◆

Suzanne has been given a page of division problems to do. She appears very frustrated and unable to write any answers. Her teacher knows that Suzanne was very successful when they worked on division with the cubes and is surprised to see how upset she is now. When the teacher goes over to talk to her, Suzanne says, "I just can't remember all these."

Suzanne is at a loss because she doesn't remember answers by rote. She does not realize that there is a way of figuring out the answers she doesn't know. Suzanne's teacher needs to help her realize that she can either imagine the cubes in her mind, get the cubes if she needs them, or draw some kind of picture to help her. Many children can do more than they think by imagining real things, but they don't use this ability because they think a number is supposed to pop into their heads.

◆ ◆ ◆

Teachers should not rush to symbols too fast or urge children to stop using materials and work abstractly before they are ready. If you take the time to build the proper foundation, the children will be much more successful later.

The Materials
and Their Preparation

The Materials and Their Preparation

Every attempt has been made to keep the materials and materials preparation as simple as possible. Black-line masters are provided for many of the task cards and game boards that you will need. Running off game boards on ditto or construction paper often works well, because they last surprisingly long and are easy to replace if they get a little worn—especially if you have run a few extra copies in the first place. Some of you will be very happy with the easy-to-make game boards and task cards, and others of you will want to take the extra time in the beginning to put them on tagboard and laminate them.

Those task cards or game boards that are specific to one game and are very easy to make are described in the same place the activity is presented. Those materials that are used over and over or are a little more complicated to make are described here.

Working Space Paper (black-line master 8):

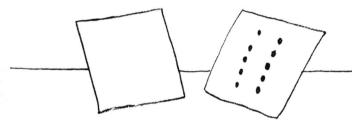

A 9 x 12 piece of paper that is used to indicate each child's individual working area. One side is blank; the other side has two rows of five dots. Using black-line master 8, run off one paper for each child in the group on light-colored construction paper or tagboard.

Counting Boards (black-line masters 9–13):

A set of counting boards consists of eight 6 x 9 cards that represent various settings. The children use cubes with these boards to represent whatever may be found in the settings.

Black-line masters 9 to 13 include the following settings, listed in the left-hand column. Ideas for objects the cubes can represent are listed in the right-hand column.

Setting	Objects
A tree 	Apples, cherries, lemons, leaves
A road 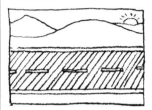	Cars, trucks, motorcycles, people, parade
A corral or pasture 	Cows, horses, pigs, cowboys
Grass (yard or field) 	Bugs, children playing, flowers, butterflies
A garden 	Vegetables, flowers
Water (ocean, pool, lake) 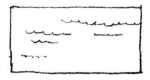	Boats, fish, shells, whales, children swimming

Barn

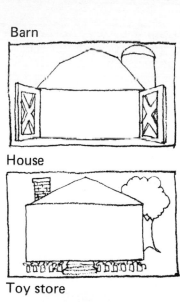

Farm animals

House

People, things in the house

Toy store

Various toys

A cave

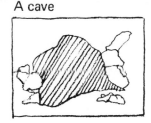

Bears, monsters, bugs, bats

Run off eight of each setting on tagboard. Color and laminate them if desired. Additional settings can be created in the following manner.

The sky (blue paper)

Birds, airplanes, bugs

Night or outer space (black paper)

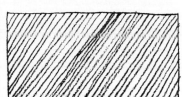

Witches, ghosts, pumpkins, stars, spaceships

Beach (sandpaper)

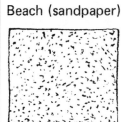

Shells, children playing, rocks, crabs

When the counting boards are used in a teacher-directed activity, each child needs one board. If you have more than eight children in a group, run off extra copies so each child in the group can have the same boards. When children are working with the counting boards independently, each child needs a complete set of eight boards.

The counting boards can be stored in clear, heavy-duty plastic bags or in pocket-type envelopes.

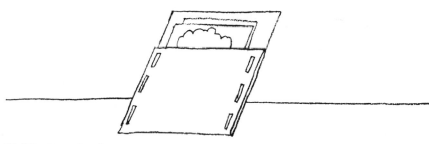

Unifix Puzzles (black-line masters 14–16):

These are a variety of shapes into which fit various amounts of Unifix cubes. The puzzles on black-line master 14 hold from three to six cubes. The puzzles on black-line masters 15 and 16 hold from seven to ten puzzles. You can distinguish between these two levels if you cut out the three-to-six puzzles from one color paper and the seven-to-ten puzzles from a second color paper. Use either construction paper or tagboard. You can cut out more than one of each puzzle if you lay the paper on which the puzzles have been run off onto a second blank piece of paper and cut both at once.

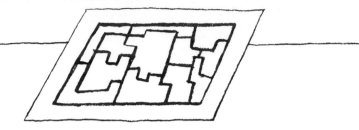

Line Puzzles (black-line masters 17–23):

These are task cards with lines of various lengths along which various numbers of cubes can be placed.

Children can place three to six cubes along the lines on black-line masters 17, 18, and 19. Seven to ten cubes can be placed along the lines on black-line masters 20, 21, 22, and 23. You can distinguish between the two levels if you run off the three-to-six line puzzles on one color paper and the seven-to-ten line puzzles on a second color paper. Cut them apart so there is one puzzle on each card.

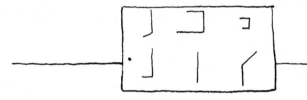

More/Less/Same Cards (black-line master 29):

Run the black-line master onto tagboard. Cut the tagboard apart to make separate more, less, and same cards.

Number Shapes (black-line masters 34-40):

The number shapes are distinctive arrangements of squares for each number from four to ten, which can be filled with Unifix cubes in various ways to show number combinations.

Using black-line masters 34–40, run the number shapes off on tagboard or construction paper, and cut them apart. For many of the activities you will need one of the number shape being worked with for each child in the group. For other activities, each child will need eight to ten of one shape.

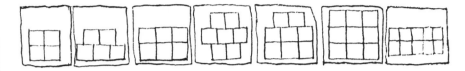

Number Train Outlines (black-line masters 41–48):

The number train outlines are drawn the same size as the Unifix cubes so that when the cubes are snapped together, they fit into the outlines exactly.

Using black-line masters 41–48, run the number train outlines off on tagboard or construction paper, and cut them apart. For many of the activities you will need one of the number train outlines being worked with for each child in the group. For other activities, each child will need eight to ten of one outline.

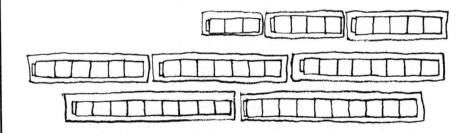

Place Value Boards

These are pieces of paper divided into two sections. Groups of cubes are placed in one section and loose cubes on the other section.

Staple or glue a 6 x 9 piece of blue paper to a 9 x 12 piece of white paper or tagboard. Draw a tiny happy face in the upper right-hand corner so the children will be able to position the board properly by making sure the happy face is right-side up.

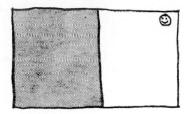

Creation Task Cards (black-line masters 26–28):

Run off the creations using black-line masters 26–28. If you want them to be on the same size cards, cut them out and mount them on tagboard.

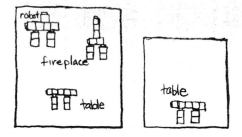

Magic Box

Cut a slit across the top of a half-gallon milk carton about half an inch deep and another slit across the bottom.

Slip a 3 x 12 piece of tagboard through the top slit, and push it out through the bottom slit. (This is easier to manage if you open up the milk carton so you can put your hand inside it. When you finish, you can staple the top back together again.) Tape the strip at the top to hold it in place.

Pull the strip out an inch or so at the bottom, and tape it in place underneath the strip. Test your magic box by placing a card inside the top slit and see if it slides down the strip easily.

in   out

Magic Box Cards

Make sets of cards to use with the magic box that have the number relationships you want your children to experience. Each set should include eight to twelve cards. Color-code each set of cards so you can keep them sorted out. For example, all the plus-one cards could be written in green, all the minus-two cards could be written in blue, etc. Make all the numbers on the backs of the cards (the answers) in black. The children will then learn to put the colored side in first to see what number will come out.

The following are examples of cards you could make.

plus one	front	1	2	3	etc.
	back	2	3	4	
minus one	front	1	2	3	etc.
	back	0	1	2	
minus two	front	2	3	4	etc.
	back	0	1	2	
plus two	front	1	2	3	etc.
	back	3	4	5	
plus three	front	1	2	3	etc.
	back	4	5	6	

Numeral Cards (see black-line masters 24, 25):

The numeral cards have dots and numerals on one side and numerals only on the other side. Children will use the cards at the level they need for success.

Run off on tagboard about twenty sheets of black-line master 24. Turn those sheets over and run black-line master 25 on the other side. Cut apart.

There are some games where the children need to find certain numerals quickly and easily. The numerals can be separated and stored in half-pint milk cartons that have been cut off and stapled together.

Addition Equation Cards (black-line masters 55–59):

Run off on tagboard and cut apart. In order to distinguish the levels, run off the sums through six on plain tag and the sums from seven to ten on a second color tagboard (or lightly shade in the paper with the side of a crayon if no colored tagboard is available).

first color	Horizontal addition, sums to six—black-line master 55
	Vertical addition, sums through six—black-line master 57
second color	Horizontal addition, sums from seven to nine—black-line master 56
	Vertical addition, sums from seven to nine—black-line master 58
	Vertical and horizontal addition, sums of ten—black-line master 59

Subtraction Equation Cards (black-line masters 60–64):

Run off on tagboard and cut apart. In order to distinguish the levels, run off differences from six on one color paper and differences from ten on a second color paper.

first color	Horizontal subtraction, differences from six and less—black-line master 60
	Vertical subtraction, differences from six and less—black-line master 62
second color	Horizontal subtraction, differences from seven to nine—black-line master 61
	Vertical subtraction differences from seven to nine—black-line master 63
	Horizontal and vertical subtraction, differences from ten—black-line master 64

Multiplication Cards (black-line masters 65, 66):

Run off on tagboard and cut apart. Because the focus is on the *process* of multiplication, not all multiplication facts are included.

Division Equation Cards (black-line master 67):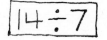

Run off on tagboard and cut apart. Because the focus is on understanding the *process* of division, not all division facts are included.

00-99 Chart (black-line master 75):

To make 00-99 charts to use with groups of children, use black-line master 75 to make a transparency, or make a large chart on tagboard and cover it with acetate. (Large sheets of acetate are available in many office supply stores.)

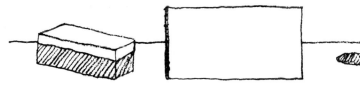

Individual Chalkboards:

Inexpensive and lightweight chalkboards can be made by cutting chipboard (available from paper companies) into the desired size (9 x 12 works nicely for the activities in this book). Paint the chipboard with chalkboard paint, which is available in many paint and hardware stores. Soft chalk is more satisfactory with these boards than hard school chalk that tends to scratch the surfaces. Have the children bring old socks for erasers, or cut regular chalkboard erasers into thirds.

Square Template:

Gather clear plastic lids such as coffee can lids. Cut out a square to match the Unifix cube (¾″ x ¾″).

Dice:

You can vary the level of difficulty for many of the activities by providing a variety of dice. Dice can be made from plain wooden cubes, pieces of foam, or commercial dice covered with blank stickers. Color-code the dots and numerals so you can distinguish between the dice quickly and easily. For example, the 0-4 dice can be made with a green pen and the 4-9 dice with a red pen, the 1-5 with a blue pen, etc.

The activities in this book call for the following dice.

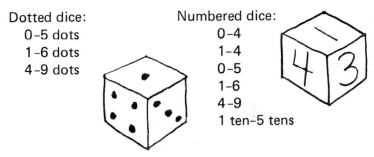

Dotted dice:	Numbered dice:
0-5 dots	0-4
1-6 dots	1-4
4-9 dots	0-5
	1-6
	4-9
	1 ten-5 tens

Large Die:

You can use a half-gallon milk carton to make a large die that can be seen by all the children in a group. Cut down three sides of the carton 3¾ inches from the bottom. Cut off the fourth side 7½ inches from the bottom.

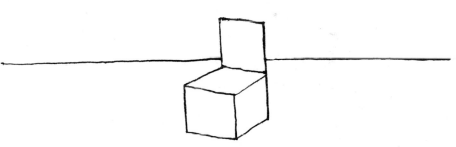

Fold down the long side and tape it.

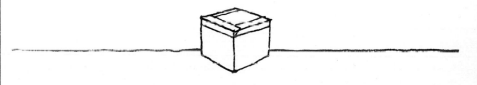

Cover the carton with plain-colored contact paper, and make the desired number of dots or numerals.

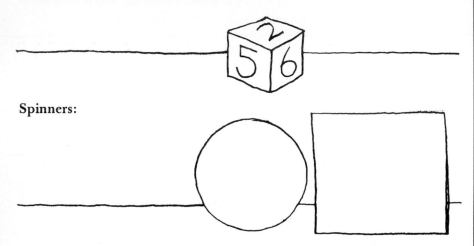

Spinners:

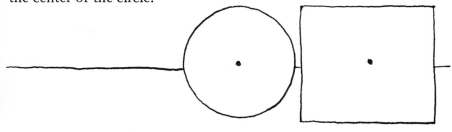

Cut a circle out of tagboard. (The circle can be made by tracing around a large margarine tub lid; it will be about six inches in diameter.) Cut a square out of tagboard slightly larger than the circle. Make the appropriate symbols on the circle according to the type of spinner you are making (see below). Poke a hole through the center of the square and the center of the circle.

Draw a line from one corner of the square to the center. This line will serve as the pointer.

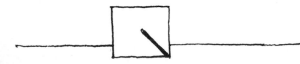

Cut out three little squares (about 1½ inches) out of the scraps of tagboard. Poke holes through the center of each and crimp by pinching them.

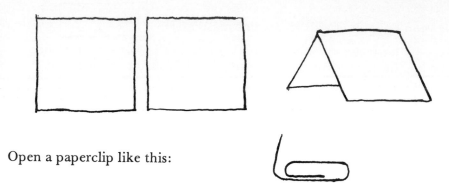

Open a paperclip like this:

Put the paperclip through the square, the three small squares, and the circle.

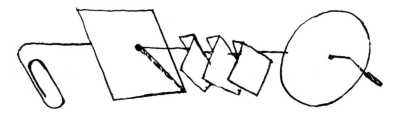

To keep the spinner from flying off and to make the end of the paperclip less sharp, wrap a tiny piece of masking tape around the end of the paperclip, making sure you do not tape it so low it touches the circle and keeps the spinner from spinning. Turn the spinner over and tape the base of the clip to the square to secure it.

Kinds of Spinners:

Plus and Minus Spinner:
Draw a line through the center of the circle. Draw a plus sign on one side of the line and a minus sign on the other.

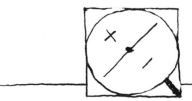

More/Less Spinner:

Draw a line through the center of the circle. Glue down the symbols for more and less from black-line master 29.

One-to-Ten Spinner

Draw lines to divide the spinner into ten equal spaces. Write the numbers from one to ten on each of the spaces.

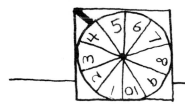

Unifix Number Lines (black-line master 32):

Fill in the number line sequence on the lines. Mount on tagboard.

Commercially Available Materials

Unifix Cubes: Many of the activities can be done with a small group of children with a set of five hundred cubes. You will be able to make greater use of the cubes if you have a set of at least one thousand.

Unifix Numeral Indicators

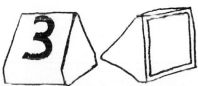

Unifix 1–10 Stair

Unifix Album of Gummed Stickers

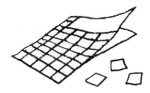

Unifix Corner Units

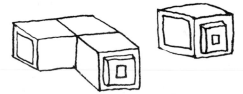

Unifix Wax Crayons

Unifix Life-Size Structured Bars Rubber Stamps

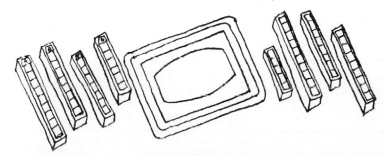

Unifix Structured Bars Rubber Stamps (smaller than life size)

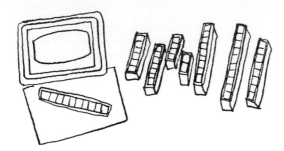

Unifix 100 Track

Unifix Multiplication/Division Markers

Unifix Value Boats

Student Worksheets

The student worksheets are open-ended, so are designed to be used over and over again for the same activity or for several different activities. Run off as many copies of each black-line master as you need, and keep a file of worksheets ready to be pulled when you need them.

Blackline Masters

1 2 3 4 5 6 7 8 9 10

Name

red	red
yellow	yellow
green	green
orange	orange
black	black
brown	brown
blue	blue
white	white
dark blue	dark blue
maroon	maroon

1	2	3	4	5	6

4	5	6	7	8	9

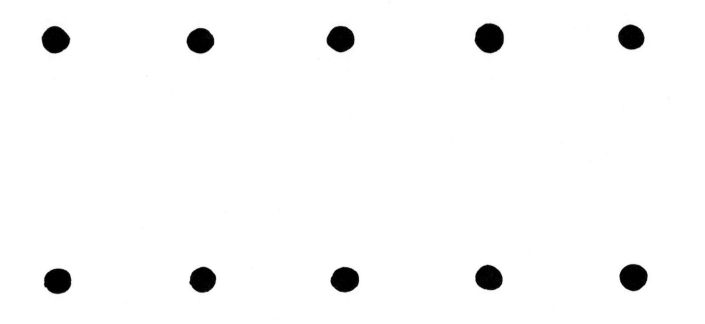

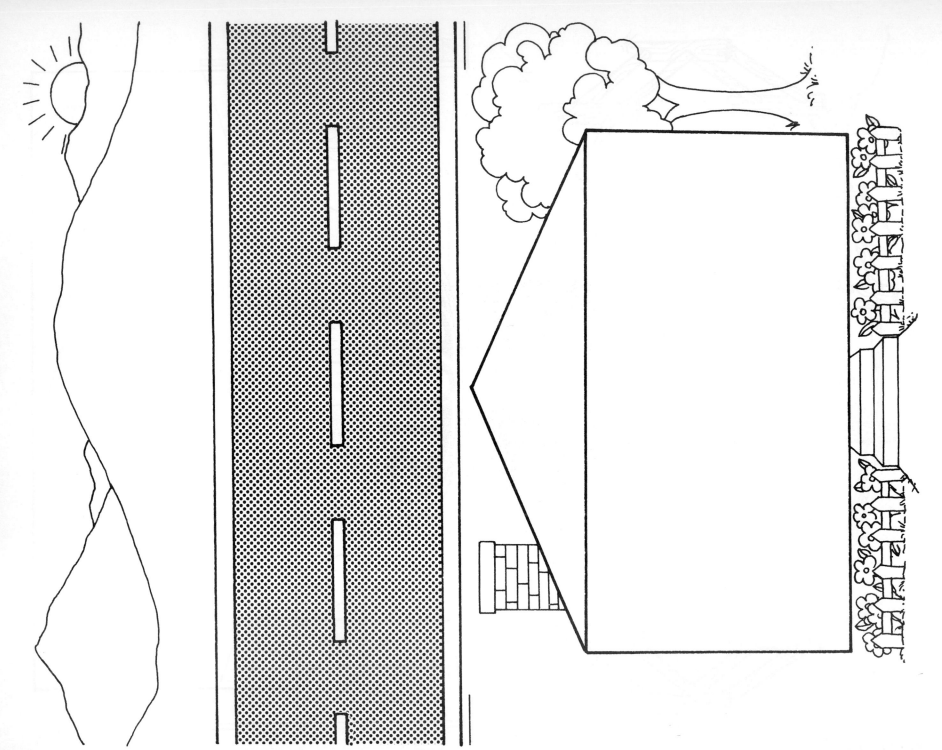

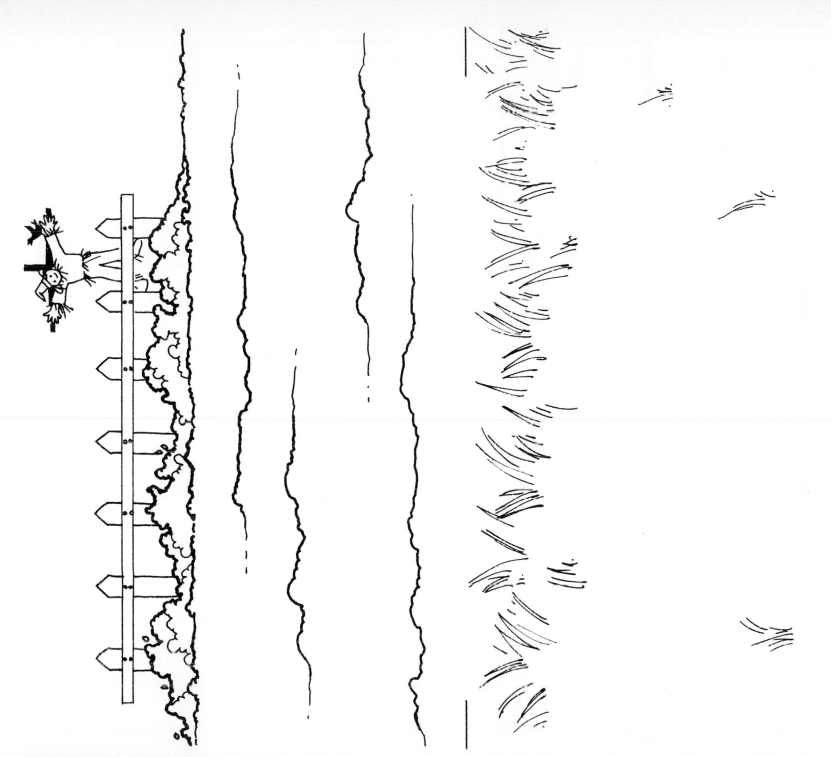

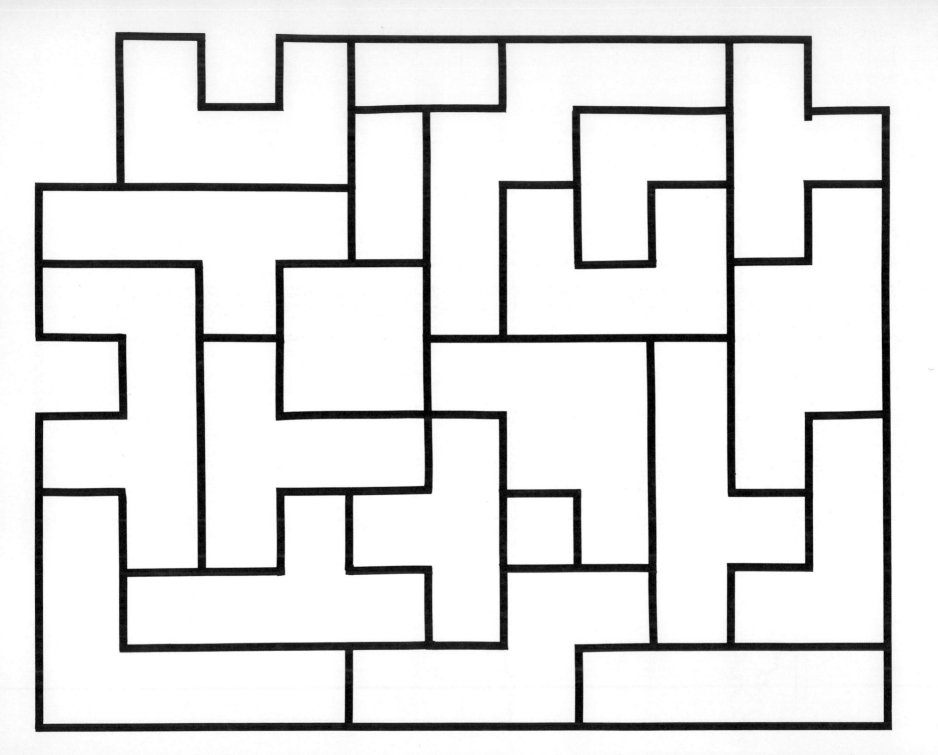

14

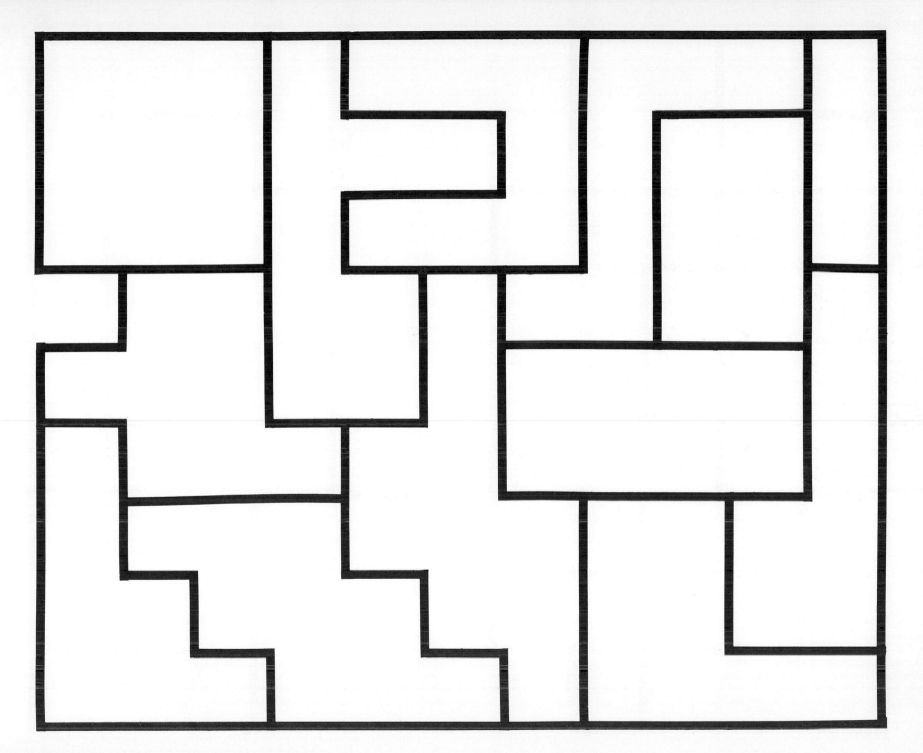

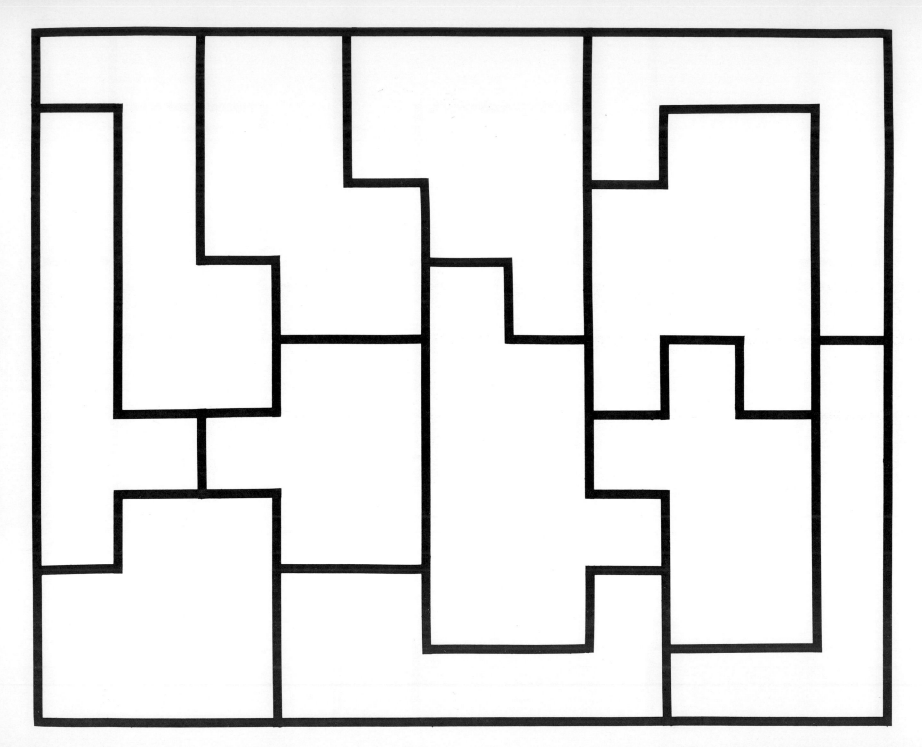

16

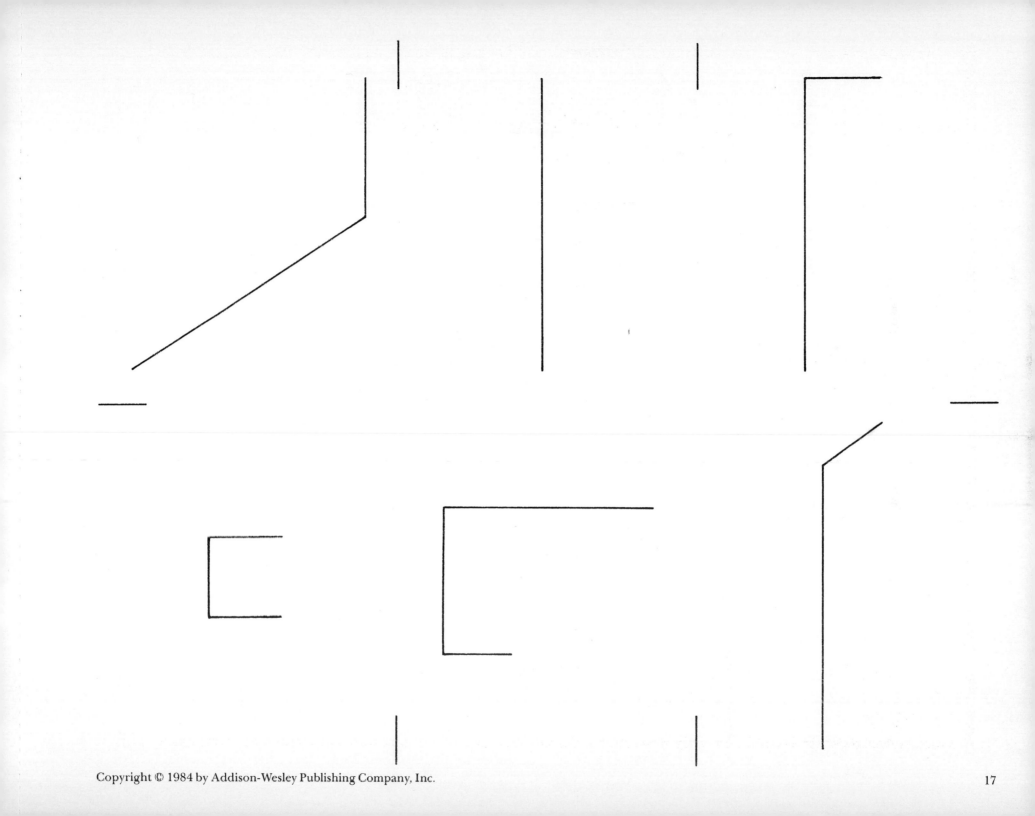

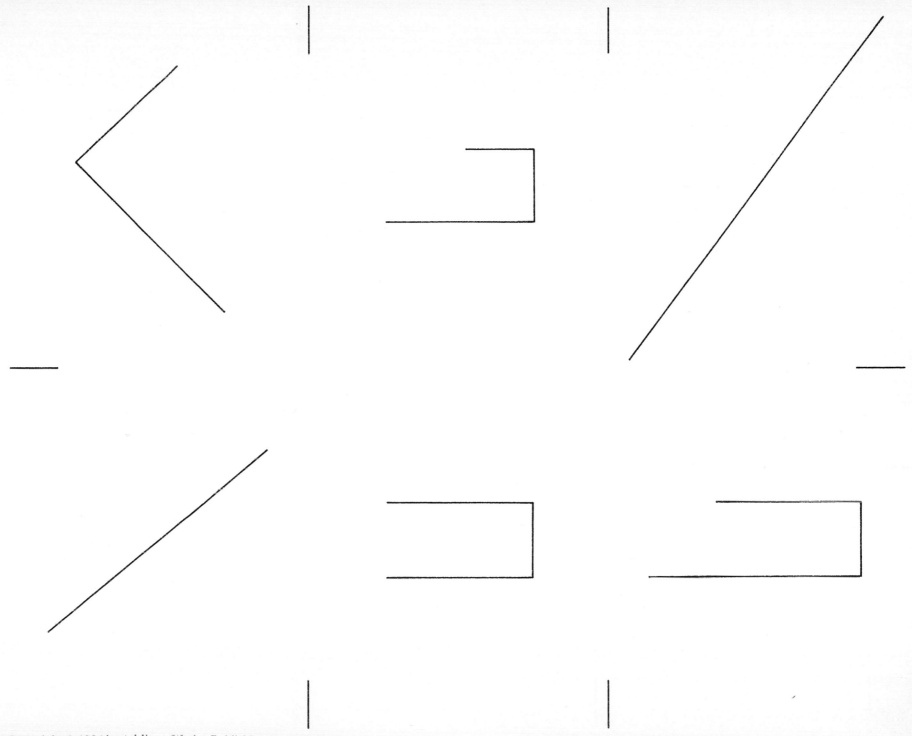

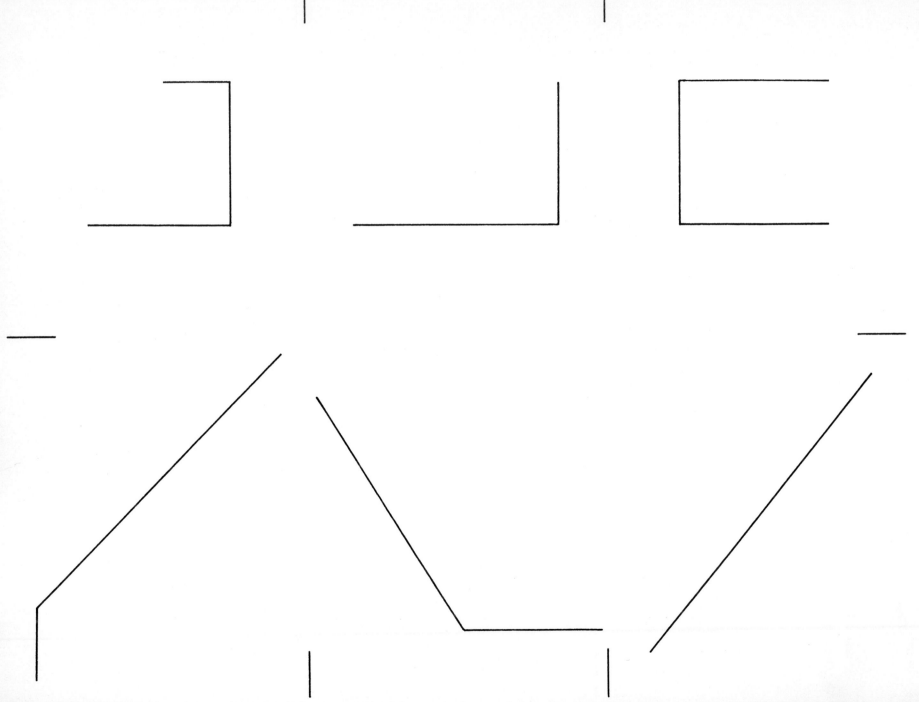

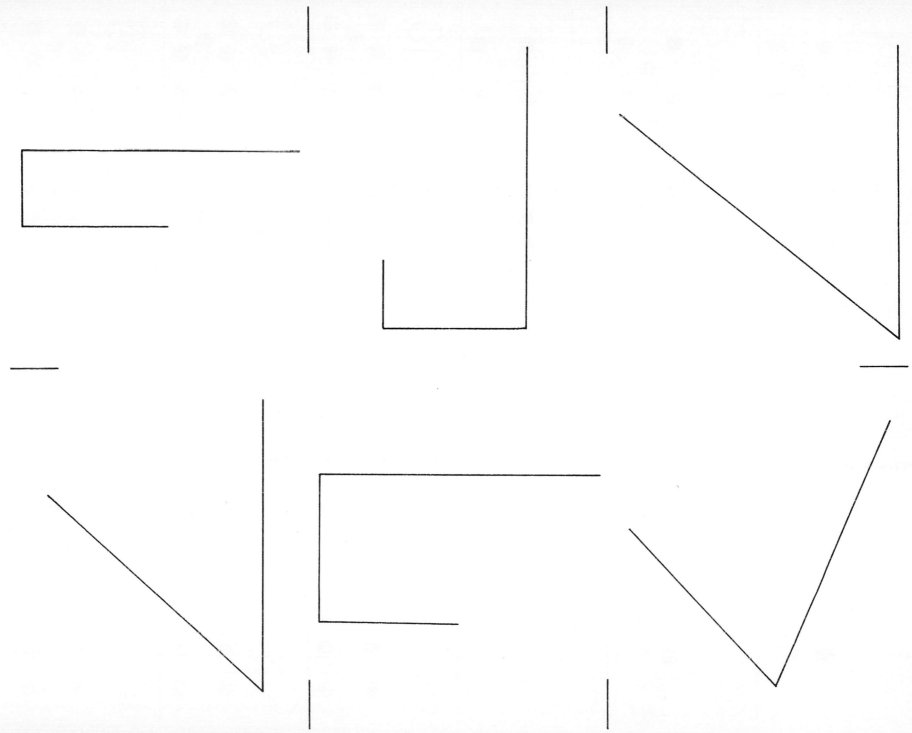

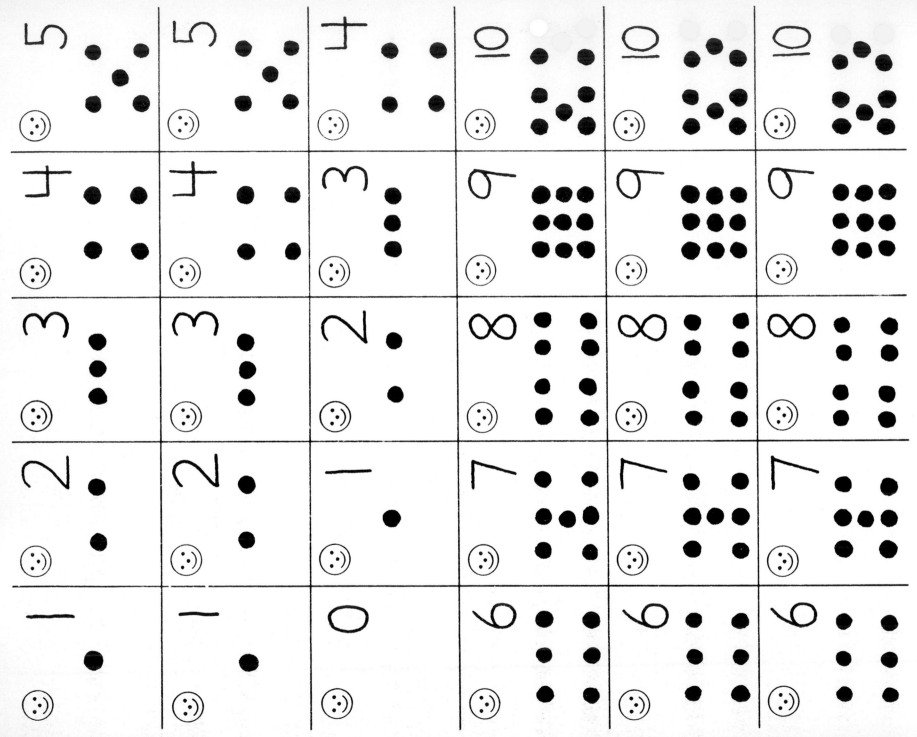

1	1	0	6	6	6
2	2	1	7	7	7
3	3	2	8	8	8
4	4	3	9	9	9
5	5	4	10	10	10

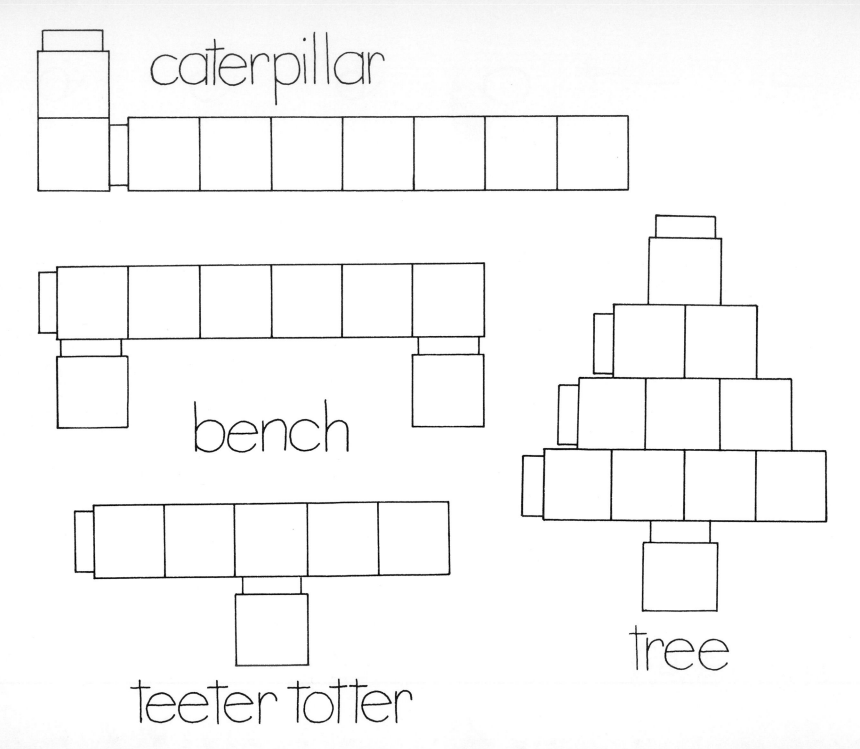

caterpillar

bench

teeter totter

tree

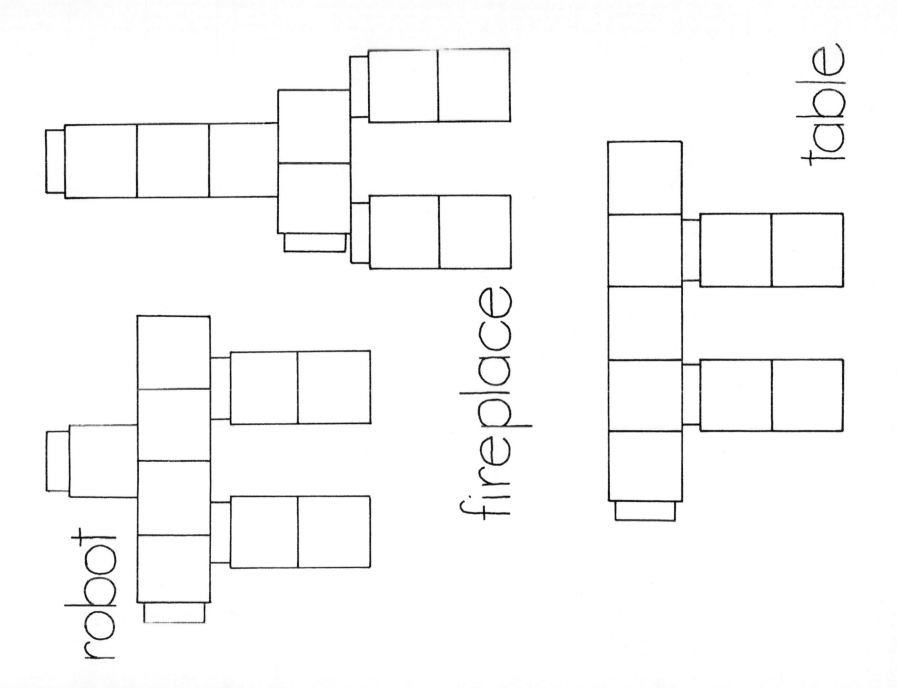

robot

fireplace

table

27

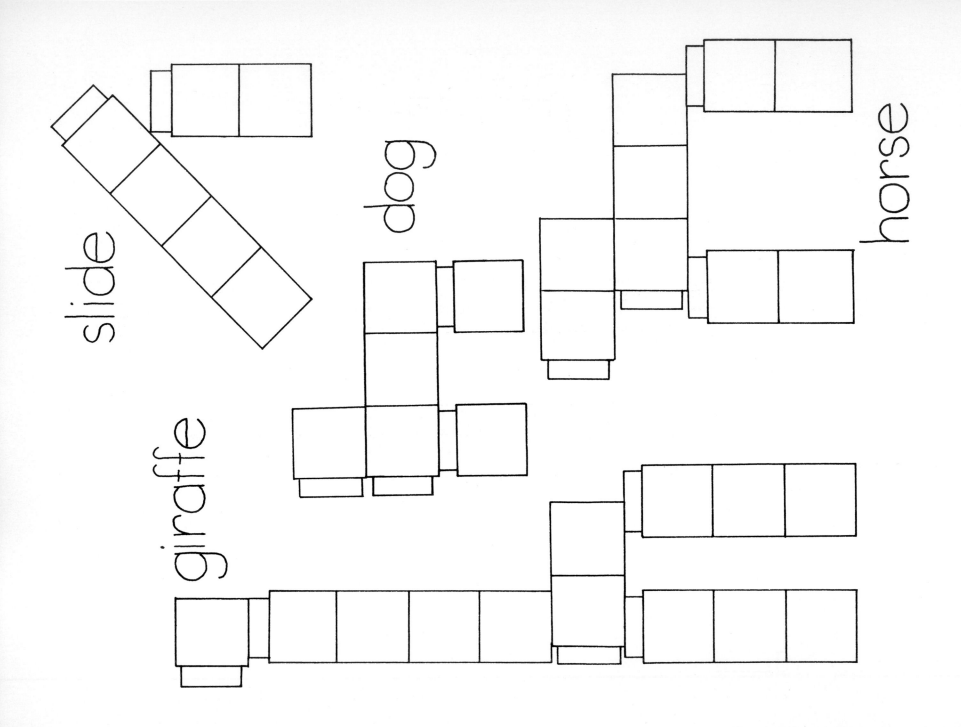

slide

dog

horse

giraffe

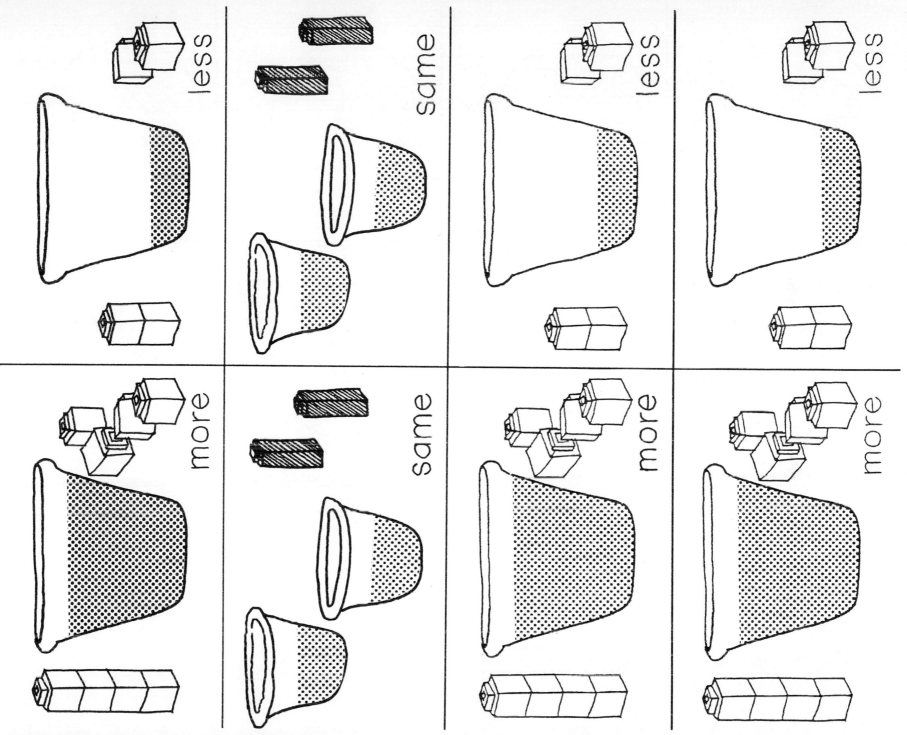

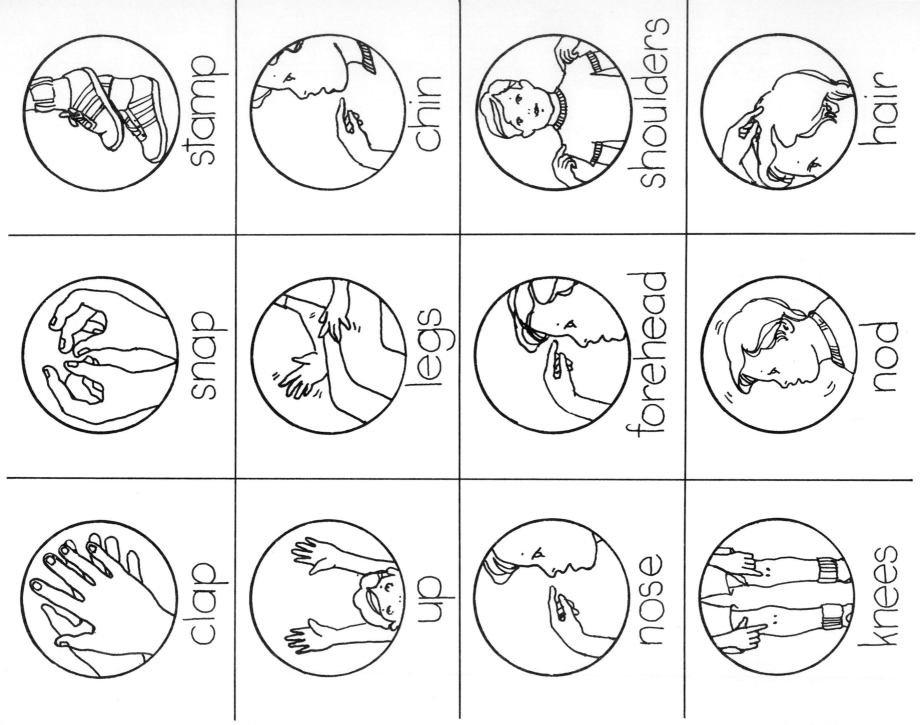

stamp

chin

shoulders

hair

snap

legs

forehead

nod

clap

up

nose

knees

Name _____

Is it more or less?

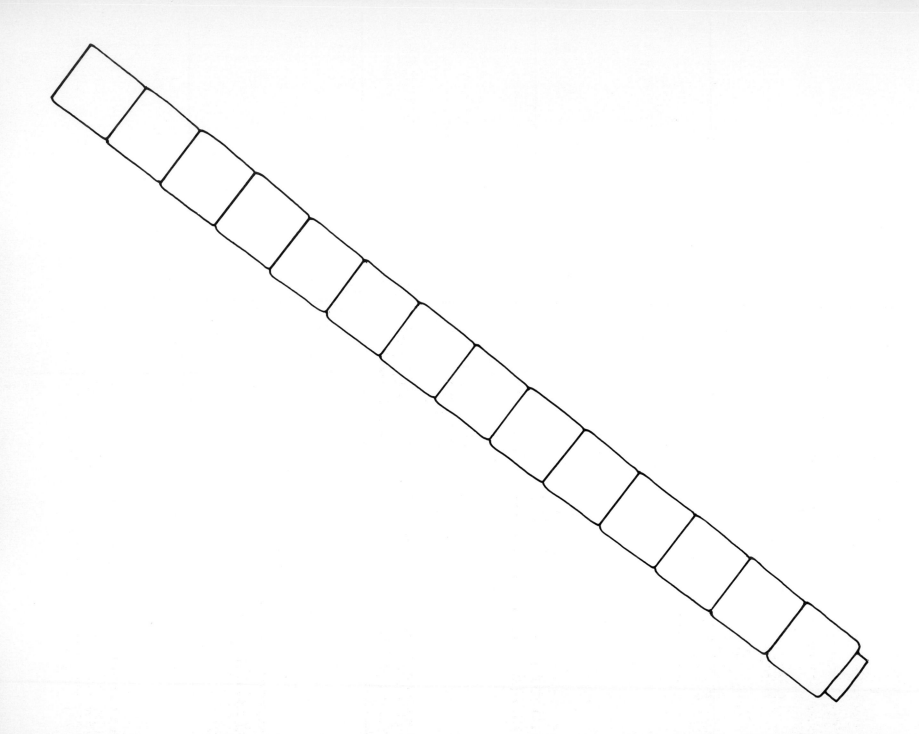

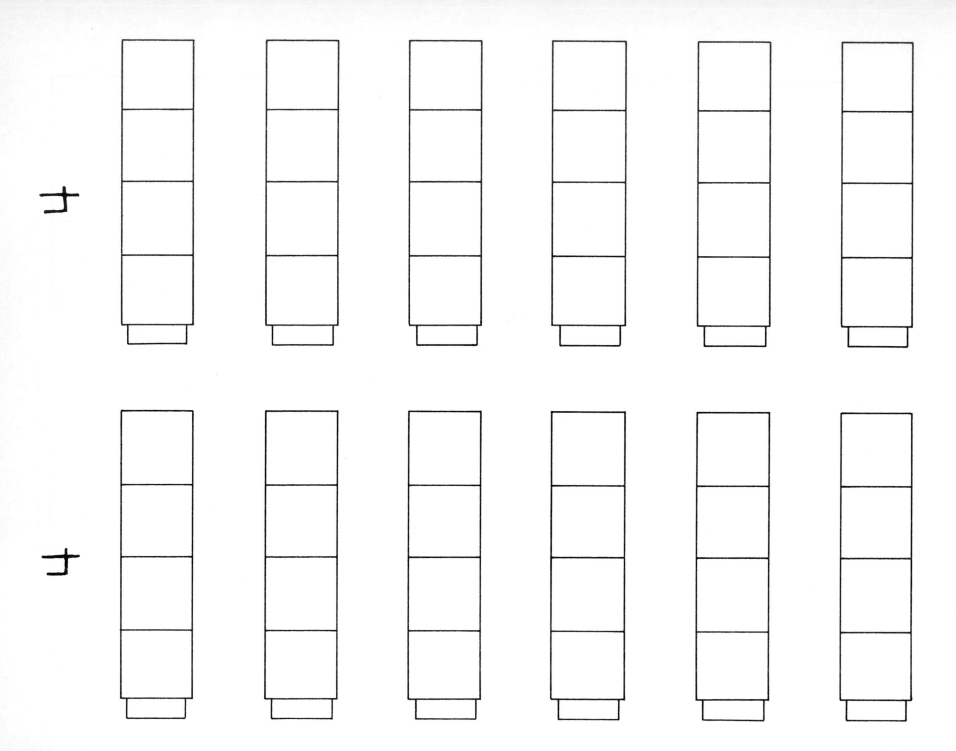

5　　　　　　　　　　　　5

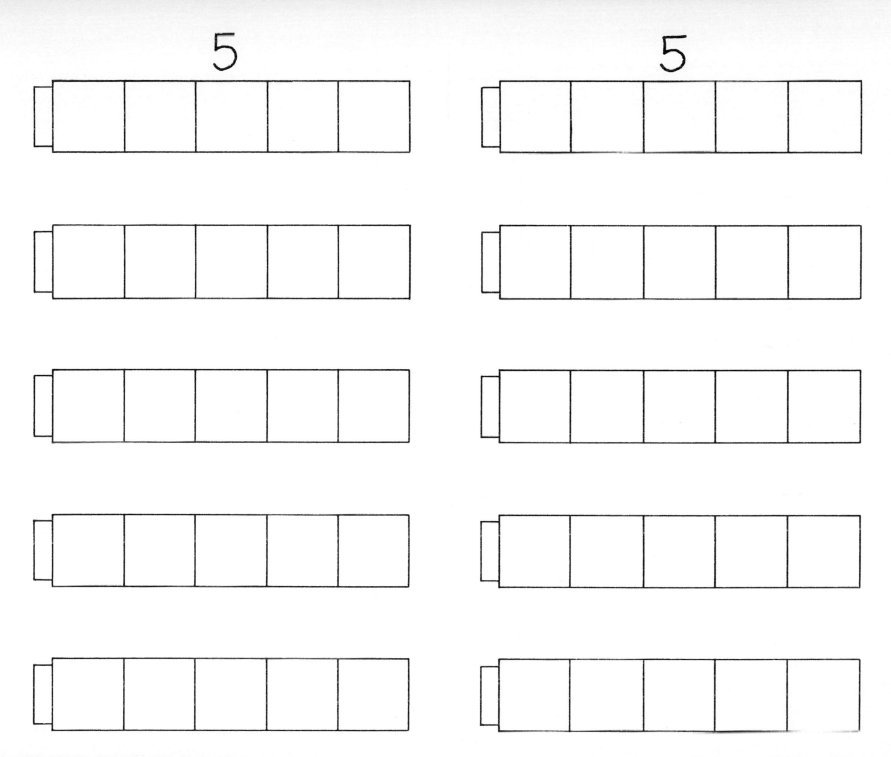

6

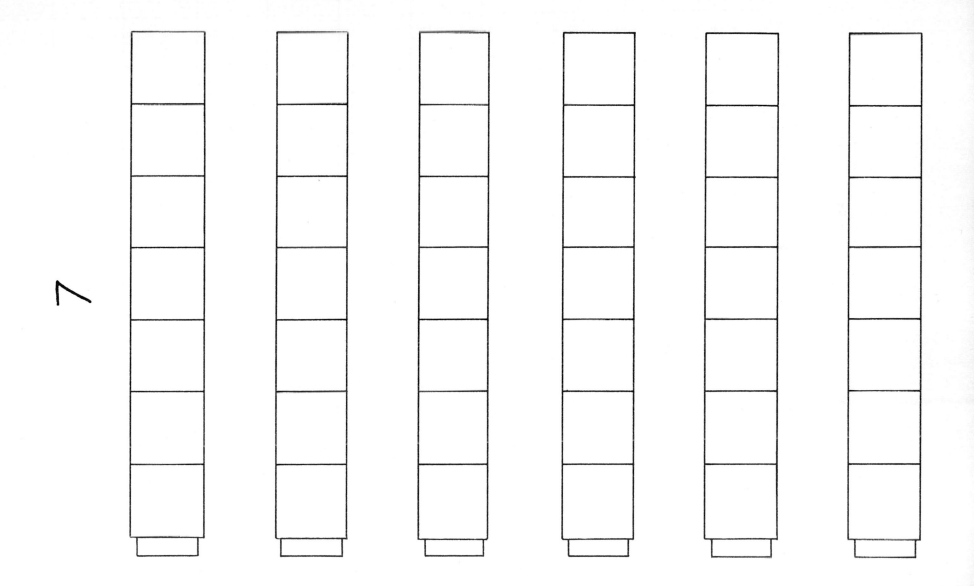

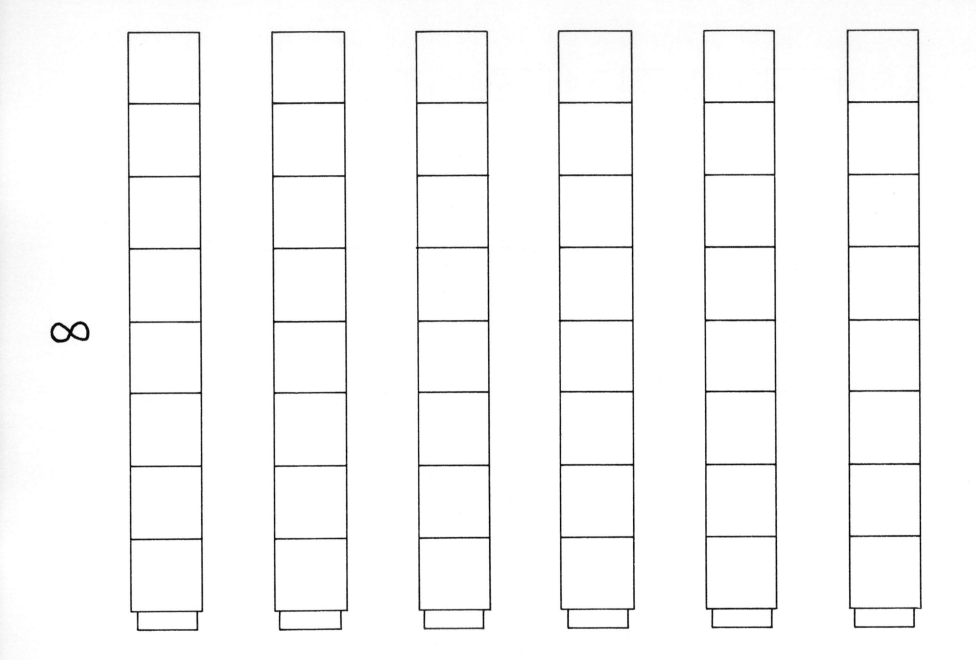

9

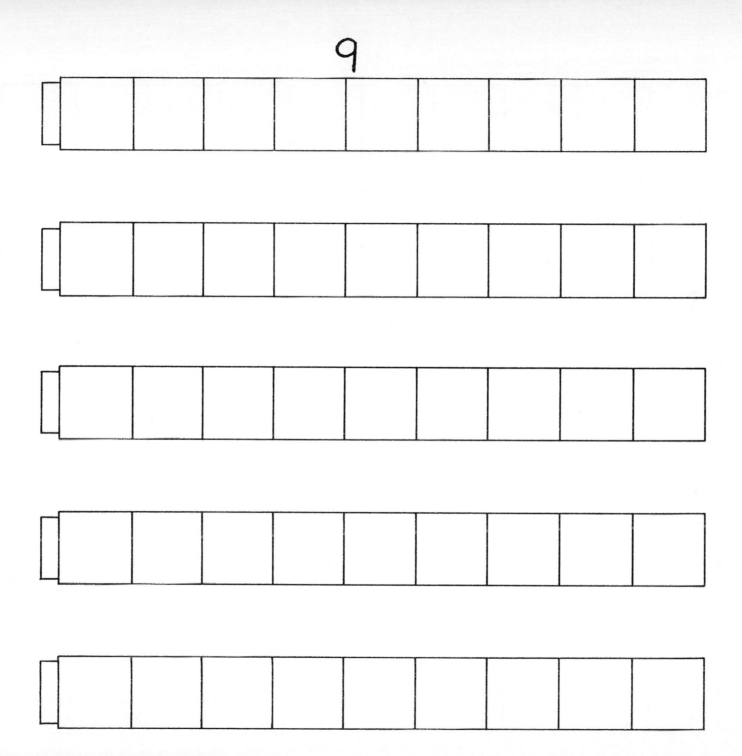

10

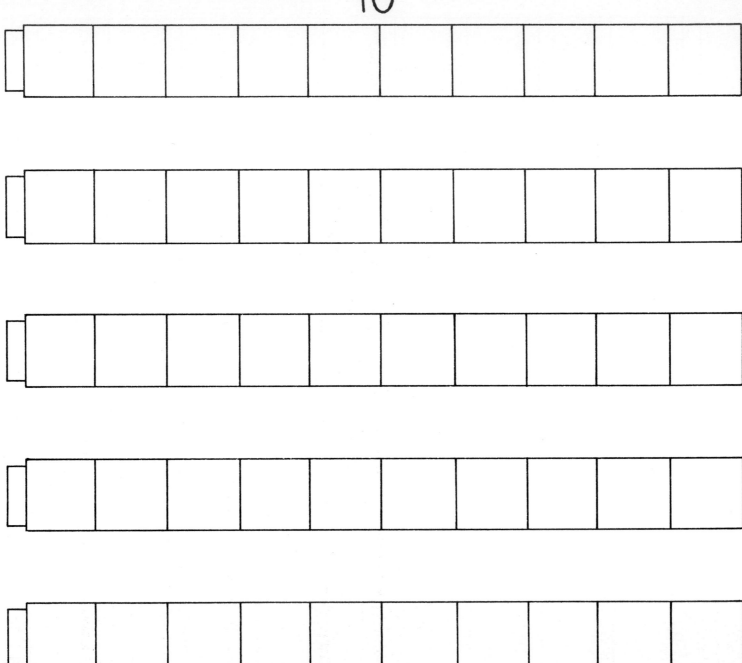

My floor is ____ cubes long.

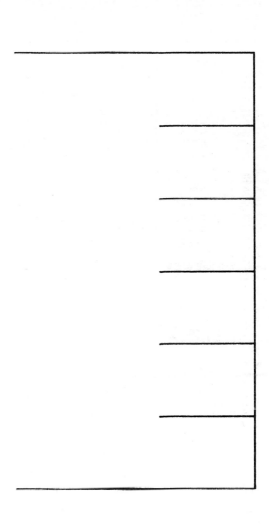

Name _____

I played a game today.
These are the equations I made.

Name _____

I worked with this number shape.

Name _____

I worked with this number train.

3

4

5

6

7

8

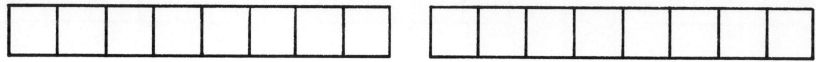

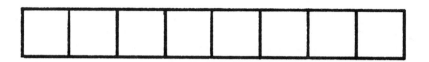

9

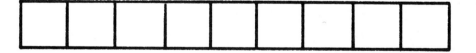

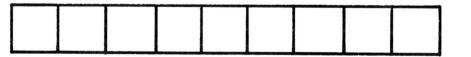

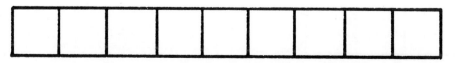

1 + 4	4 + 0	1 + 0
0 + 5	3 + 1	0 + 1
6 + 0	2 + 2	2 + 0
5 + 1	1 + 2	1 + 1
4 + 2	0 + 4	0 + 2
3 + 3	5 + 0	3 + 0
2 + 4	4 + 1	2 + 1
1 + 5	3 + 2	1 + 2
0 + 6	2 + 3	0 + 3

1 +0	0 +1	2 +0	1 +1	0 +2	3 +0	0 +3	1 +2	2 +1
4 +0	0 +4	1 +3	3 +1	2 +2	5 +0	4 +1	3 +2	2 +3
1 +4	0 +5	6 +0	1 +5	2 +4	3 +3	4 +2	5 +1	0 +6

8 + 1	7 + 1	7 + 0
7 + 2	6 + 2	6 + 1
6 + 3	5 + 3	5 + 2
5 + 4	4 + 4	4 + 3
4 + 5	3 + 5	3 + 4
3 + 6	2 + 6	2 + 5
2 + 7	1 + 7	1 + 6
1 + 8	0 + 8	0 + 7
0 + 9	9 + 0	8 + 0

7 +0	6 +1	5 +2	4 +3	3 +4	2 +5	1 +6	0 +7	8 +0
7 +1	6 +2	5 +3	4 +4	3 +5	2 +6	1 +7	0 +8	9 +0
8 +1	7 +2	6 +3	5 +4	4 +5	3 +6	2 +7	1 +8	0 +9

4 +6	3 +7	2 +8	1 +9	0 +10	8 +2	4 +6	1 +9	3 +7
2 + 8	1 + 9	0 + 10	10 +0	9 +1	2 +8	7 +3	6 +4	5 +5
10 + 0	9 + 1	8 + 2	7 + 3	6 + 4	5 + 5	4 + 6	3 + 7	7 + 3

5 − 4	5 − 5	6 − 1	6 − 2	6 − 3	6 − 4	6 − 5	6 − 6	6 − 0
4 − 0	4 − 1	4 − 2	4 − 3	4 − 4	5 − 0	5 − 1	5 − 2	5 − 3
1 − 0	2 − 0	2 − 1	1 − 1	2 − 2	3 − 0	3 − 1	3 − 2	3 − 3

1	1	2	2	2	3	3	3	3
-0	-1	-0	-1	-2	-0	-1	-2	-3

4	4	4	4	4	5	5	5	5
-0	-1	-2	-3	-4	-0	-1	-2	-3

5	5	6	6	6	6	6	6	6
-4	-5	-0	-1	-2	-3	-4	-5	-6

9 − 1	9 − 2	9 − 3	9 − 4	9 − 5	9 − 6	9 − 7	9 − 8	9 − 9
8 − 1	8 − 2	8 − 3	8 − 4	8 − 5	8 − 6	8 − 7	8 − 8	9 − 0
7 − 0	7 − 1	7 − 2	7 − 3	7 − 4	7 − 5	7 − 6	7 − 7	8 − 0

7	8	9	8	8	8	9	9	9
-0	-1	-1	-2	-3	-6	-0	-4	-7
7	7	7	7	8	8	9	9	9
-1	-3	-5	-7	-4	-7	-2	-5	-8
7	7	7	8	8	8	9	9	9
-2	-4	-6	-0	-5	-8	-3	-6	-9

10 −7	10 −8	10 −9	10 −10	10 −2	10 −4	10 −6	10 −7	10 −8
10−9	10−10	10 −0	10 −1	10 −2	10 −3	10 −4	10 −5	10 −6
10−0	10−1	10−2	10−3	10−4	10−5	10−6	10−7	10−8

1×4 1	1×6 1	1×7 1
6×1 6	5×1 5	2×1 2
0×3 0	0×7 0	0×4 0
2×0 2	3×0 3	6×0 6
2×2 2	2×3 2	2×4 2
2×5 2	2×6 2	2×7 2
3×2 3	3×3 3	3×4 3
4×3 4	3×5 3	3×6 3
4×2 4	4×5 4	4×6 4

5 × 4	5 × 3	5 × 2
2 × 8	3 × 8	3 × 7
6 × 2	5 × 6	5 × 5
6 × 5	6 × 4	6 × 3
1 × 5	1 × 3	1 × 2
4 × 7	4 × 4	2 × 9
8 × 2	7 × 3	7 × 2
9 × 3	9 × 2	8 × 3
0 × 9	9 × 1	1 × 9

$6 \div 1$	$6 \div 2$	$6 \div 3$
$7 \div 3$	$7 \div 2$	$6 \div 4$
$8 \div 4$	$8 \div 3$	$8 \div 2$
$9 \div 4$	$9 \div 2$	$9 \div 3$
$12 \div 6$	$12 \div 4$	$12 \div 3$
$14 \div 2$	$14 \div 7$	$12 \div 5$
$15 \div 5$	$15 \div 2$	$15 \div 3$
$20 \div 3$	$20 \div 4$	$15 \div 4$
$25 \div 4$	$20 \div 2$	$20 \div 5$

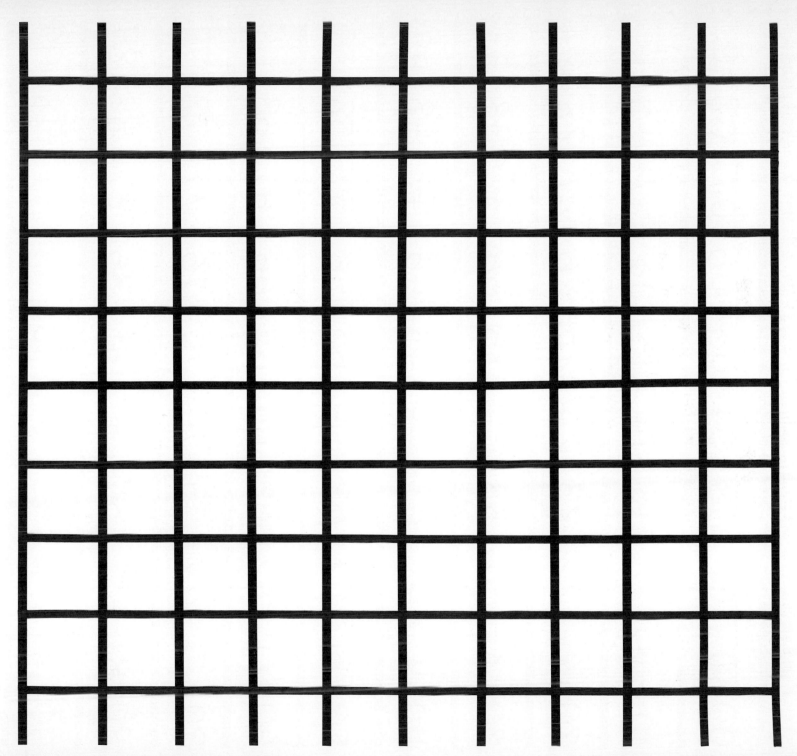

09	19	29	39	49	59	69	79	89	99
08	18	28	38	48	58	68	78	88	98
07	17	27	37	47	57	67	77	87	97
06	16	26	36	46	56	66	76	86	
05	15	25	35	45	55	65	75	85	95
04	14	24	34	44	54	64	74	84	94
03	13	23	33	43	53	63	73	83	93
02	12	22	32	42	52	62	72	82	92
01	11	21	31	41	51	61	71	81	91
00	10	20	30	40	50	60	70	80	90

70

Name _____

I worked with _____

My guess		I found out	
tens	ones	tens	ones

☐ ☐ ☐ ☐

Name _____

I am _____ unifix cubes long.

longer than me	shorter than me

tens ones | tens ones | tens ones | tens ones

4 0   2 4   3 0   1 8
+ 4   + 2   + 3   + 1
_____ _____ _____ _____

3 5   2 1   1 9   2 7
+ 3   + 2   + 1   + 2
_____ _____ _____ _____

2 6   1 4   3 2   3 6
+ 2   + 1   + 3   + 3
_____ _____ _____ _____

1 7   2 2   3 1   2 5
+ 1   + 2   + 3   + 2
_____ _____ _____ _____

73

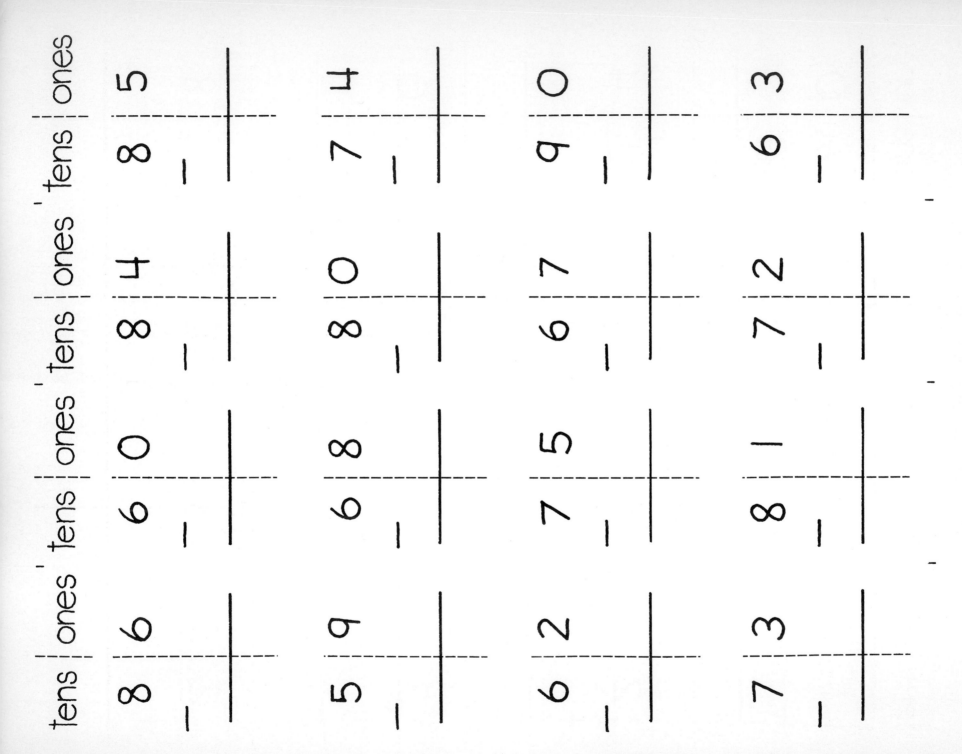

Name

How many? (cups, rows, groups, stacks)	How many in each?	How many altogether?

$0 \times$ $1 \times$ $2 \times$ $3 \times$ $4 \times$ $5 \times$ $6 \times$ $7 \times$ $8 \times$ $9 \times$

$\times 0$ $\times 1$ $\times 2$ $\times 3$ $\times 4$ $\times 5$ $\times 6$ $\times 7$ $\times 8$ $\times 9$

9 × 9   2 × 2   2 × 2

3 × 3   5 × 5   2 × 2

8 × 8   3 × 3   4 × 4

2 × 2   4 × 4   5 × 5

4 × 4   6 × 6   7 × 7

2 × 2   3 × 3   6 × 6

3 × 3   2 × 2   2 × 2

6 × 6   7 × 7   3 × 3

$\overline{)9}$	$\overline{)13}$	$\overline{)15}$	$\overline{)19}$	$\overline{)21}$	$\overline{)17}$	$\overline{)24}$	$\overline{)25}$
$\overline{)40}$	$\overline{)27}$	$\overline{)16}$	$\overline{)12}$	$\overline{)18}$	$\overline{)34}$	$\overline{)37}$	$\overline{)23}$
$\overline{)36}$	$\overline{)35}$	$\overline{)34}$	$\overline{)33}$	$\overline{)32}$	$\overline{)31}$	$\overline{)30}$	$\overline{)29}$
$\overline{)20}$	$\overline{)18}$	$\overline{)16}$	$\overline{)14}$	$\overline{)12}$	$\overline{)10}$	$\overline{)8}$	$\overline{)6}$

25 ÷ 35 ÷ 16 ÷

32 ÷ 27 ÷ 15 ÷

18 ÷ 22 ÷ 14 ÷

23 ÷ 31 ÷ 13 ÷

12 ÷ 26 ÷ 12 ÷

17 ÷ 29 ÷ 11 ÷

37 ÷ 20 ÷ 10 ÷

14 ÷ 33 ÷ 9 ÷